On the Extent
and Aims of a
National Museum
of Natural History

RICHARD OWEN

CAMBRIDGE
UNIVERSITY PRESS

CAMBRIDGE UNIVERSITY PRESS

Cambridge, New York, Melbourne, Madrid, Cape Town,
Singapore, São Paolo, Delhi, Tokyo, Mexico City

Published in the United States of America by Cambridge University Press, New York

www.cambridge.org
Information on this title: www.cambridge.org/9781108038294

© in this compilation Cambridge University Press 2011

This edition first published 1862
This digitally printed version 2011

ISBN 978-1-108-03829-4 Paperback

CAMBRIDGE LIBRARY COLLECTION

Books of enduring scholarly value

Life Sciences

Until the nineteenth century, the various subjects now known as the life sciences were regarded either as arcane studies which had little impact on ordinary daily life, or as a genteel hobby for the leisured classes. The increasing academic rigour and systematisation brought to the study of botany, zoology and other disciplines, and their adoption in university curricula, are reflected in the books reissued in this series.

On the Extent and Aims of a National Museum of Natural History

A significant limitation on the development of zoology, botany and palaeontology in the mid-nineteenth century was the absence of a centralised collection of specimens. Appointed superintendent of the British Museum's natural history departments in 1859, the distinguished biologist Richard Owen (1804–92) quickly realised the need to make various scattered samples more readily available for study, and began campaigning for a new, national museum with unprecedented space and resources. This book is the text of one of his speeches to the Royal Institution, given in 1861 and first published in 1862. He argues against the usual practice of exhibiting only one type form for each genus, provides possible floor plans, and presents case studies across the zoological field which show the limitations of the then current system. He also stresses a new idea, that such a museum should aim not only to help scientists, but to educate the general public.

Cambridge University Press has long been a pioneer in the reissuing of out-of-print titles from its own backlist, producing digital reprints of books that are still sought after by scholars and students but could not be reprinted economically using traditional technology. The Cambridge Library Collection extends this activity to a wider range of books which are still of importance to researchers and professionals, either for the source material they contain, or as landmarks in the history of their academic discipline.

Drawing from the world-renowned collections in the Cambridge University Library, and guided by the advice of experts in each subject area, Cambridge University Press is using state-of-the-art scanning machines in its own Printing House to capture the content of each book selected for inclusion. The files are processed to give a consistently clear, crisp image, and the books finished to the high quality standard for which the Press is recognised around the world. The latest print-on-demand technology ensures that the books will remain available indefinitely, and that orders for single or multiple copies can quickly be supplied.

The Cambridge Library Collection will bring back to life books of enduring scholarly value (including out-of-copyright works originally issued by other publishers) across a wide range of disciplines in the humanities and social sciences and in science and technology.

ON THE

EXTENT AND AIMS

OF

A NATIONAL MUSEUM

OF

NATURAL HISTORY.

INCLUDING

THE SUBSTANCE OF A DISCOURSE ON THAT SUBJECT, DELIVERED
AT THE ROYAL INSTITUTION OF GREAT BRITAIN, ON THE
EVENING OF FRIDAY, APRIL 26, 1861.

BY PROFESSOR OWEN, F.R.S.,

FOREIGN ASSOCIATE OF THE INSTITUTE OF FRANCE, ETC.

LONDON:

SAUNDERS, OTLEY, & CO.,

66, BROOK STREET, HANOVER SQUARE.

1862.

LONDON : PRINTED BY WILLIAM CLOWES AND SONS, STAMFORD STREET,
AND CHARING CROSS.

DESCRIPTION OF THE PLATES.

PLATE I.

PLAN of the premises surrounding the British Museum, showing the proportion of those to the west, extending from Great Russell Street, along Charlotte Street, to Bedford Square, which would be occupied by the proposed Museum of Natural History, as adapted for the present Collections.

Fig. 1. Plan of Ground Floor of the proposed Museum, as adapted to any part of the oblong plot of ground extending from Great Russell Street to Montague Place; and to front either west, or east as drawn in the plan. (The details of the galleries, &c., are described in the text.)

PLATE II.

Plan of the ground to the west of the Horticultural Garden, along Prince Albert's Road, Kensington; showing the proportion which would be occupied by the proposed Museum of Natural History, as adapted for the reception of the present Collections in the British Museum.

Fig. 1. Plan of the Ground Floor.

Fig. 2. Plan of the Upper Floor.

The dotted lines show where galleries might be added, conformably with the original design, for future accessions to the Natural History. The building is adapted to any part of the oblong plot of ground extending from the Exhibition Building towards Kensington Road, and might be erected with the entrance to face the end of Gore Road.

Fig. 3. Plan (elevation in two sections) of the Museum, as ultimately extended along a frontage of 780 feet.

The section, along A, A, *fig.* 2, through the transverse galleries, shows the mode of admitting light into the ground-floor L, from the interspaces of the galleries; and the mode of utilizing the interspaces below the side-lights of the ground-floor by an arched roof of glass. The section through the longitudinal galleries, along B, B, *fig.* 2, exposes their communications with the transverse galleries and interspaces.

"Let Mr. Owen describe exactly the kind of building that will answer his purpose—that will give space for his whales and light for his humming-birds and butterflies. The House of Commons will hardly, for very shame, give a well-digested scheme so rude a reception as it did on Monday night."— 'Times,' May 21st, *Leader on the Museum Debate.*

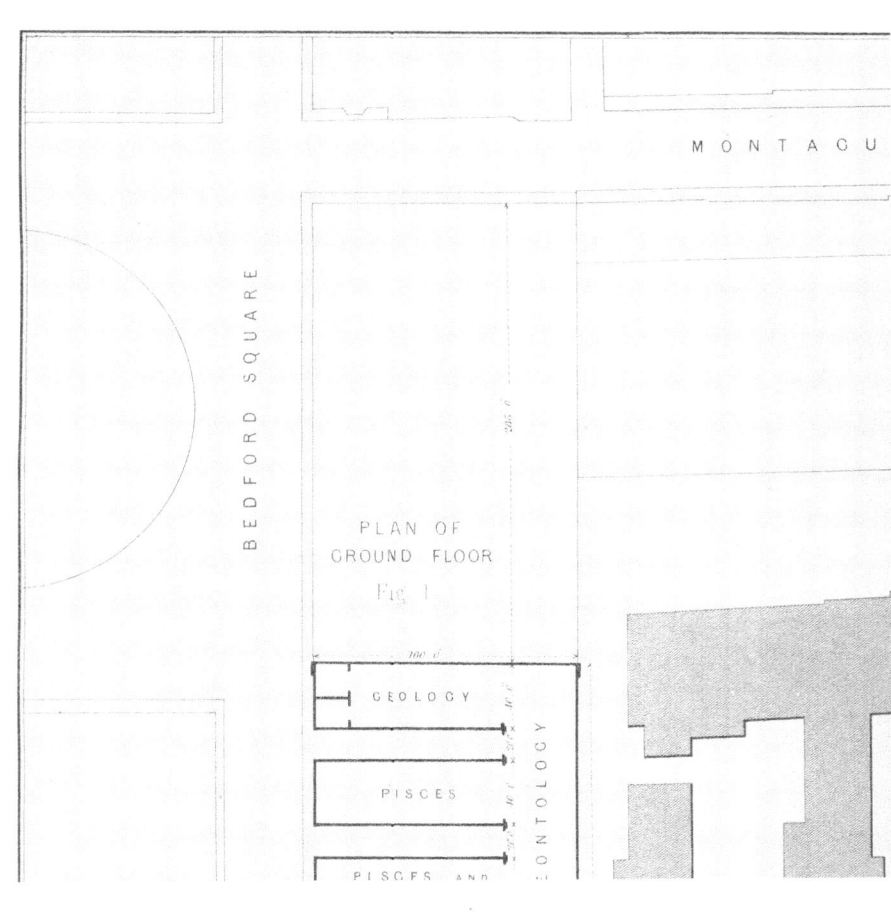

MONTAGU

BEDFORD SQUARE

PLAN OF
GROUND FLOOR

Fig. 1

GEOLOGY

PISCES

PISCES AND

PÆONTOLOGY

Pl. I.

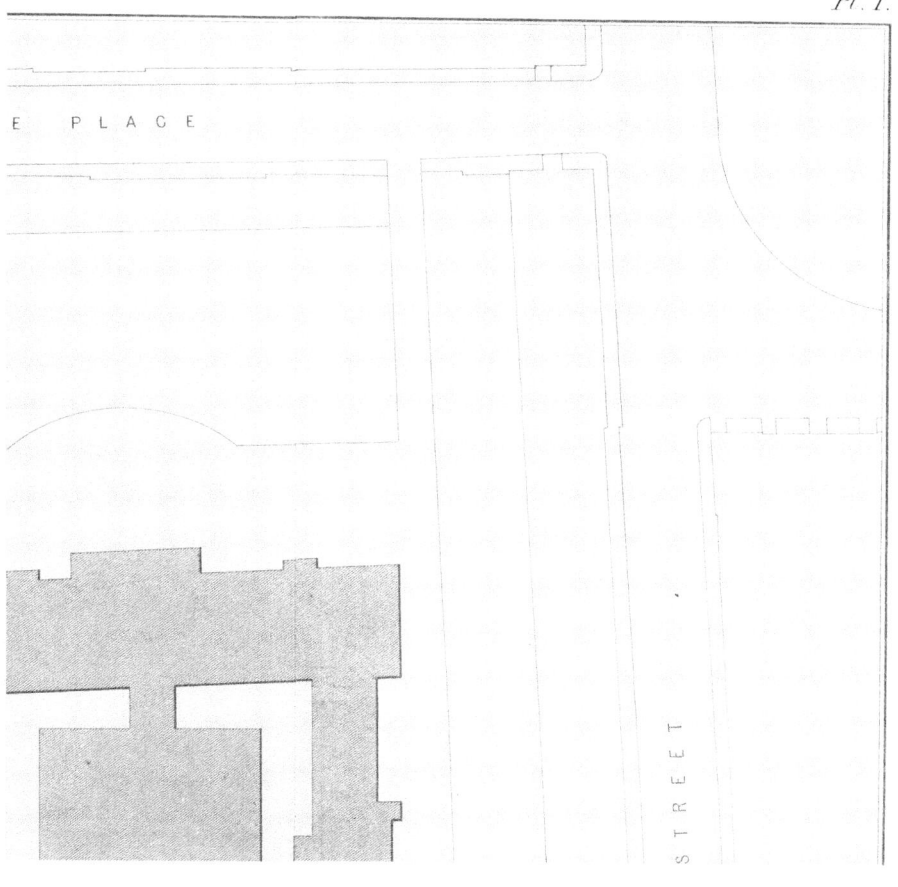

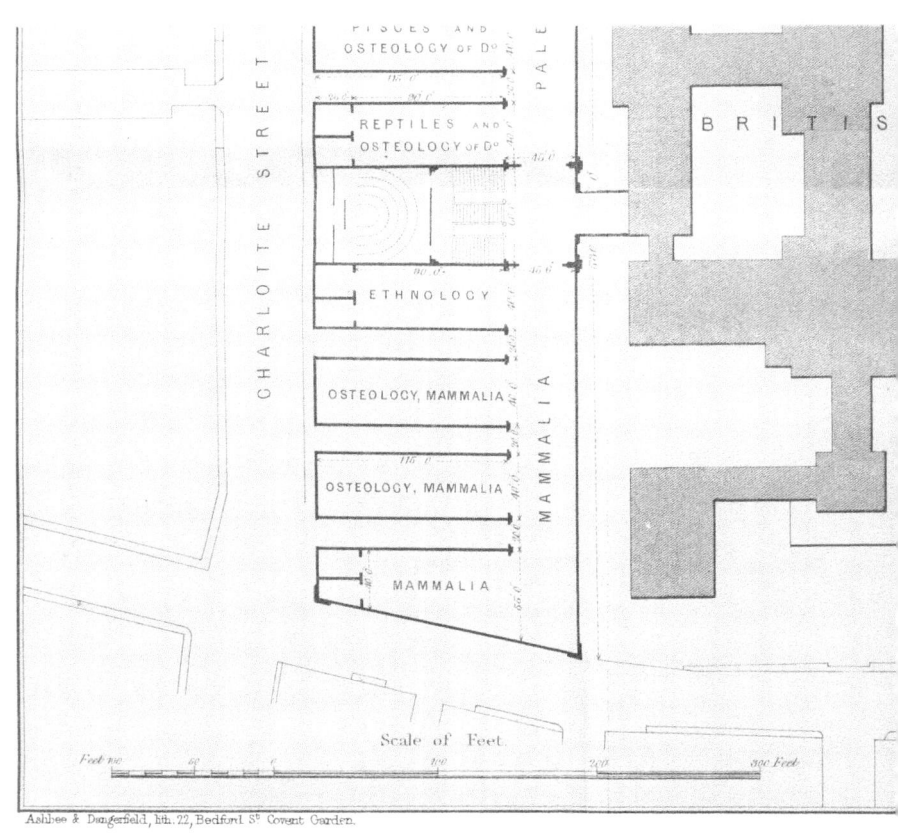

FISUES AND
OSTEOLOGY OF Dᵒ

REPTILES AND
OSTEOLOGY OF Dᵒ

ETHNOLOGY

OSTEOLOGY, MAMMALIA

OSTEOLOGY, MAMMALIA

MAMMALIA

CHARLOTTE STREET

PALE

MAMMALIA

BRITIS

Scale of Feet.

Feet 100   50   0   100   200   300 Feet

PLAN OF MUSEUM OF NATURAL HIS

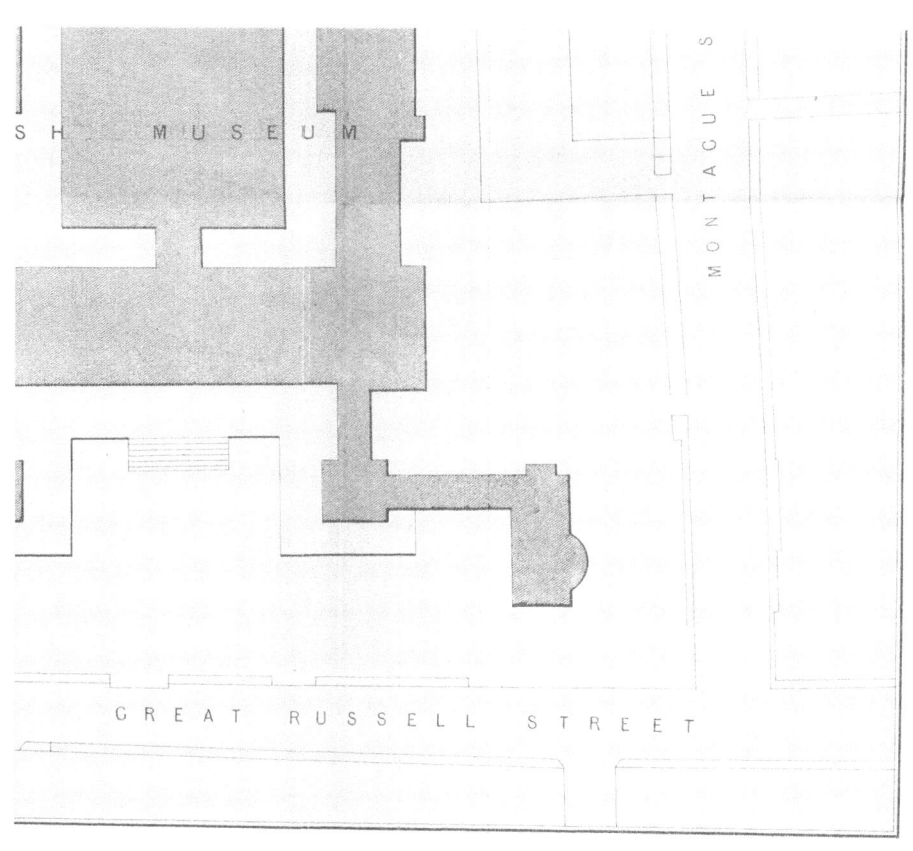

STORY (GROUND FLOOR), BLOOMSBURY.

Pl. II

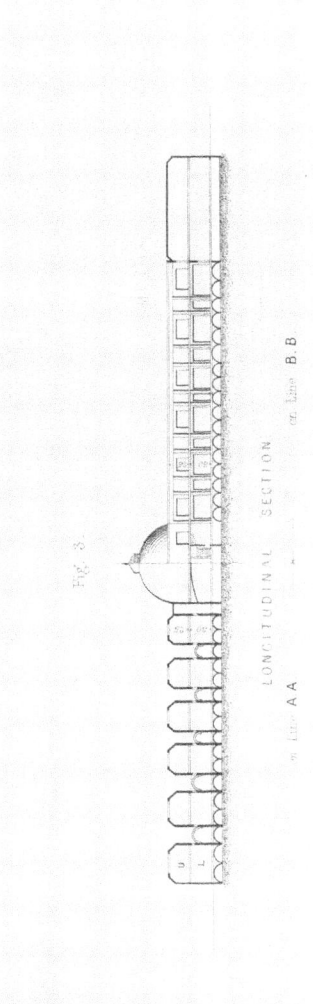

Fig. 3.

LONGITUDINAL SECTION

" line A.A — or line B.B

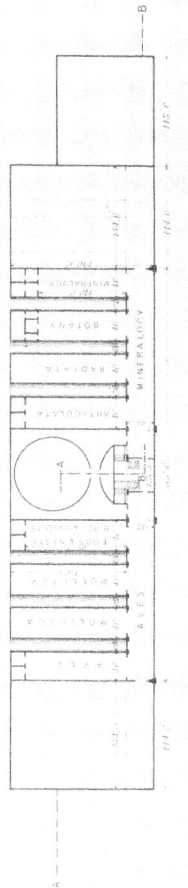

PLAN OF UPPER FLOOR

FIG. 3.

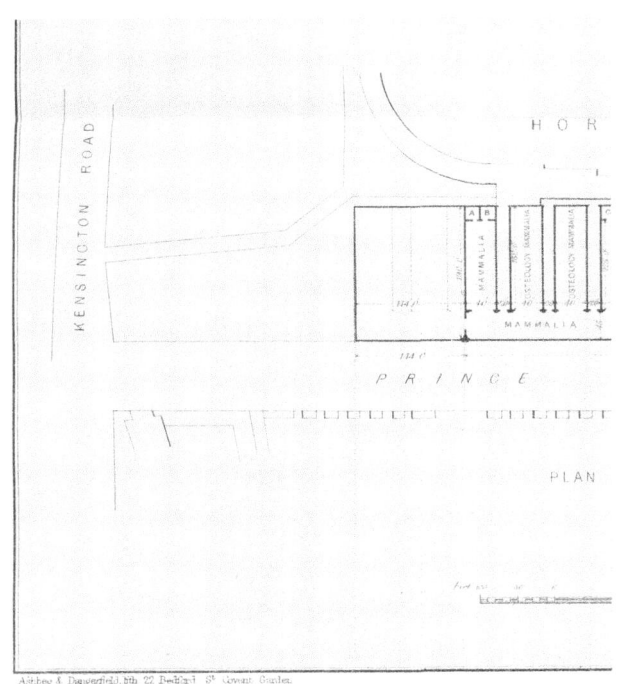

PLAN OF MUSEUM

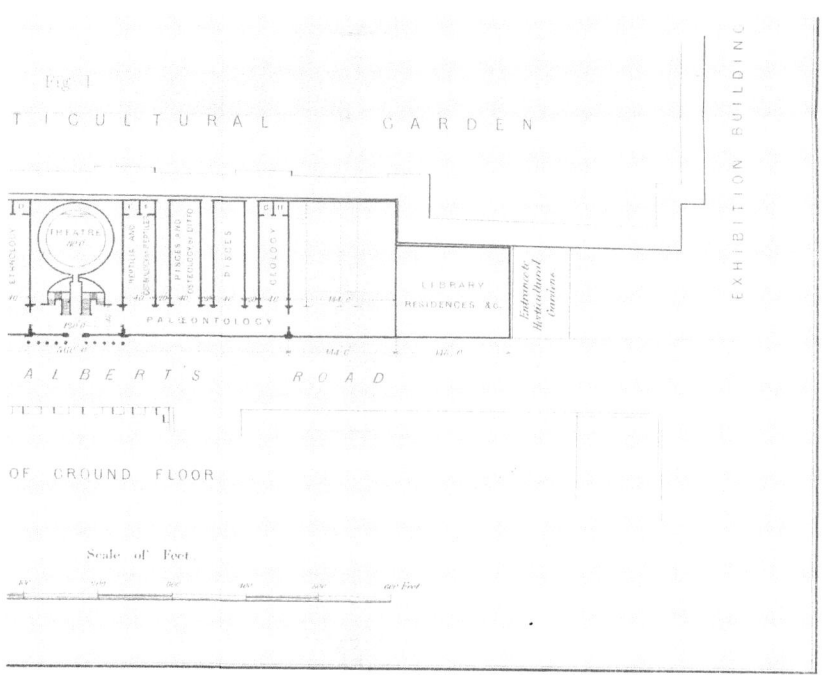

Fig. 1

TICULTURAL GARDEN

EXHIBITION BUILDING

THEATRE

ZOOLOGY

ZOOPHYTES &c

PISCES &c
BATRACIA & REPT

AVES

GEOLOGY

PALÆONTOLOGY

LIBRARY
RESIDENCES &c.

Entrance to
Horticultural
Gardens

ALBERT'S ROAD

OF GROUND FLOOR

Scale of Feet

OF NATURAL HISTORY, KENSINGTON.

ON A

NATIONAL MUSEUM

OF

NATURAL HISTORY.

MANY inquiries by those interested in the intellectual progress of the country and the material helps thereto had moved me to put the facts and arguments in behoof of an adequate National Museum of Natural History in a popular and accessible form, when the Leading Journal, divining with its usual sagacity the generality of such desire after late debates, made the appeal, the response to which will be found in the following pages.

To descant on the abstract advantage of a knowledge of the works of Creation is neither requisite nor convenient to my present purpose. I may assume the general admission that collections of the several classes of such objects, duly prepared,

named, and arranged, so as to give the utmost
facility for inspection and comparison, are the indispensable instruments in the acquisition and
advance of that knowledge. Not but that a vital
part of Natural History requires the observation
of rocks and mineral beds as they naturally
appear on the earth's surface, of plants as they
clothe and adorn that surface, of animals as they
add life and motion to sea, earth, and air. And
such knowledge must be gained abroad, in the
field, in that grand Natural Museum which the
world becomes to the loving eyes of the geological, botanical, and zoological observer, even such
as was the Paradise in which Adam, as sung by
our great poet,

> " Beheld each bird and beast
> Approaching two and two ; these cowering low
> With blandishment, each bird stooped on his wing.
> He named them as they passed and understood
> Their nature ; with such knowledge God endued
> His sudden apprehension."

Under other and harder conditions we strive
to regain that knowledge, needing, and urgently
seeking for, every collateral aid in the struggle
to acquire that most precious commodity — the

truth as it is in Nature, and as manifested by the works of God.

ARISTOTLE received such aid from his great pupil ALEXANDER, in large subventions for the requisite subjects of his numerous observations on the external form and anatomy of animals, and for the employment of hunters and fishers and other observers of their living habits. He thus obtained the materials on which his strong intellect wrought in the composition of the remarkable ' History of Animals' of which we probably form but an inadequate idea from the nine books that have come down to us. Had the Greeks, indeed, possessed and practised the arts of preserving and preparing animal bodies and structures, the science of the universal teacher might have been retained, exemplified, and expanded by his pupils. But there is no record of any collection of the con-servable parts of the animals having been made after the philosopher had used them for his obser-vations and comparisons. To the absence of such museum may be attributed the singular and sudden arrest in the course of the zoological science, which had started by so rapid a growth, but which, want-

ing that condition of progress, degenerated ; so that
even Bacon failed to comprehend the zoological
discoveries of the Stagyrite ; and their due appreciation had to await the zootomical labours of HUNTER
and CUVIER. Not until their time, and by the
aid of their vast collections of animals and animal
structures, could the value and importance of
Aristotle's zoological and physiological writings
be fully comprehended.

The Romans, during the height of their Empire,
expended enormous treasures in the capture and
transport to Italy of the wild beasts of their con-
quered provinces, but they were solely for the
service of the amphitheatre. The rhinoceros, the
hippopotamus, the giraffe, the crocodile, the lion,
the tiger, the European bison, with many other
rare animals—some of which, as the gigantic wild
ox of Hyrcania, are now extinct—were brought to
Rome, and there publicly exhibited. They might
be seen by the philosopher, the historian, the poet,
and the satirist; but they were seen only to be
baited and slaughtered in cruel games for the
gratification of the depraved tastes of an enslaved
and voluptuous people.

Many centuries elapsed ere the nations of Europe began again to be familiar with the forms of the rarer animals of remote regions; and we may regard it as one of the beneficial results of the great moral revolution which had been effected during that interval, that Natural History objects began to be collected through other and higher motives than those which stimulated the Pagans of ancient Rome to excel in the exhibitions of the circus.

The Sophists and Epicureans of the Empire had ample opportunities during more than 400 years of making observations on the form and organization of foreign animals; but it seems that these animals, after being slaughtered, were put to no further use.

We may justly rejoice that, in these later times, aims engendered and supported by Christian principles, under a sense of responsibility for the use of talents and opportunities, have stimulated states and communities to bring together collections of objects from the different kingdoms of Nature, and to arrange them in orderly series, with a view to minister to the advancement of science and to the instruction, elevation of thought, and innocent pleasures of the peoples.

Every European nation now possesses its National Museum of Natural History, aiming, or professing, to be a more or less complete epitome of the three kingdoms of Nature—Animals, Plants, and Minerals. Such is the extent and scope of the great Natural History Museum or establishment at the Jardin des Plantes in Paris'; and such also, in its fundamentals, is the Natural History Division of the British Museum.

Here the Natural History Departments are four in number : one for Zoology, comprising the prepared specimens, skeletons, shells, &c., of existing species of animals ; one for Geology, at present restricted to the exhibition of the remains of extinct organised beings ; a third for Botany, and a fourth for Mineralogy : so that there needs only a co-ordinate or consistently proportional display of the several classes of objects in the four great Departments of this comprehensive design, to realise our ideal of the scope of a National Museum of Natural History.

These several Departments of Natural History

are so related to each other as to give and receive
reciprocal illustration, and to make it requisite for
the student of any one class of objects to have
ready access to the rest : the Galleries of Zoology,
Botany, Geology, and Mineralogy, should therefore
form part of one building. The Department of
Zoology in such a Museum should be so located, in
reference to the Department of Palæontology, as to
afford the easiest transit from the specimens of
existing to those of extinct animals. The Geologist
specially devoted to the study of the evidences of
extinct vegetation, ought, in like manner, to have
means of comparing his fossils with the collections
of recent plants.

The Department of Zoology, considered in its
ordinary restricted application to the present phase
of living nature, should illustrate both the outward
form and inward structure of its subjects. The
art of Taxidermy preserves the integument, so
as to give the shape and exemplify the diverse
coverings of mammals, birds, reptiles, fishes,
insects, crustaceans, and echinoderms. In a great
proportion of the Molluscous classes, the external
defensive shell is all that is commonly seen of the

living animal; and consequently the shell itself represents in a great measure the outward form of such. Where the soft parts of the body are habitually protruded from the shell in the ordinary living actions, they can be closely imitated in coloured wax models. Such models, or the Mollusks themselves in transparent colourless antiseptic liquor, will complete the representation of the outward forms of the species in this great department of Invertebrate Animals.

Most of the Arachnida (spiders, mites), the larvæ of the Insecta (caterpillars, maggots), many Crustacea, some of the Cirripedia (soft-stalked barnacles), all the Annelida (sea centipedes, nereids, tubeworms, earthworms, leeches) and Entozoa (intestinal and other internally parasitic worms, flukes, and hydatids) can only be exhibited either preserved in liquor or by means of models.

In the Radiated classes the testa, with its appendages, serves to convey a sufficient idea of the external form of the animal in the Sea-urchin and Starfish orders. The corals, madrepores, and other hard parts of the Polypes, are the mere skeletons of the animals: these masses of carbonate of lime are generally internal, rarely visible in the living colony; and as the delicate flesh shrinks in most preserving liquors, and loses colour in all, a true idea of these beautiful composite marine animals, with the polype-mouths expanded like the petals of a flower, can only be given by coloured models. The art of the colourist and modeller is still more essential to show the external characters of the lovely hyaline medusæ, jelly-fish, and other forms of the floating Acalephes; for only the form, and that commonly contracted, can be studied in specimens preserved in liquor.

The adjustment of the exquisitely sculptured silicious shells of microscopic organisms to microscopes adapted to afford the casual visitor to a Public Museum the opportunity of contemplating them, must be restricted to a few of the most striking examples. Enlarged coloured drawings, with the degree of the magnifying power noted, may serve to exemplify the majority of the species. A special apartment, with an attendant accustomed to the use of the microscope, would be requisite in the Public Museum, in which the general view of living Nature was completed by a demonstration of the forms of the infusorial and other animalcules.

The most obvious and simple aim of the Zoological Department of a National Museum of Natural History is to exhibit the various outward forms and characters of the animal kingdom; and this, as we have just seen, requires the labours of skilled artists in various costly procedures. But the first principle in the arrangement and allocation of such objects is, that each class of animals should receive its due proportional amount of elucidation to the extent which the acquired specimens at the time may admit, and according to the degree in which the principle of variety is manifested in the class.

A museum of Nature does not aim, like one of Art, merely to charm the eye and gratify the sense of beauty and of grace. Many animal forms do indeed accord with our apprehension of the Beautiful; some classes more especially, as e. g. that of birds, also the pearly shells in which mollusks "attend soft nutriment," the diversely-ramified or delicately-sculptured corals—all these are strikingly beautiful, and accordingly are the exemplifications of animated Nature which are the first to be collected, and are usually the most extensively illustrated in museums. But there are forms of animals which excite wonder by their bulk and power, which surprise by their strangeness and oddity, which repel by their ugliness, or excite an instinctive feeling of horror and disgust. How soon the latter emotions subside, as the adaptation of form and structure to the habits and exigencies of the species become understood, the observant Naturalist can readily testify; and it would be difficult indeed to say what exceptional proportion of the animal kingdom may not possess that element of beauty which rests in the appreciation of the perfect fitness of the thing to its function. As, however, the purpose of a Museum of Natural History is to set forth the extent and variety of the Creative Power, with the sole rational aim of imparting and diffusing that knowledge which begets the right spirit in which all Nature should be viewed, there ought to be no partiality for any particular class merely on account of the quality which catches and pleases the passing gaze.

The first and great essential for a co-equal or

justly proportional representation of all the classes of animals to the extent in which a nation may possess, or have opportunities to acquire, the specimens of them, is adequate exhibition-space. This is to be estimated—first, by the number of known species of the class; secondly, by the extent of exhibition-space occupied by the proportion of the class which may be properly exhibited in any existing museum; thirdly, by the proportion of examples obtained, but not exhibited in such museum; fourthly, by the ratio at which such specimens have accrued in a given number of years, and by the circumstances or conditions on which the ratio of future increase may be computed; finally, by the proportion of the specimens required to be exhibited in orderly series for a true idea of the group, and for the convenience of the student.

MAMMALIA.—Let me first exemplify the working of this problem by the present class.

In the year 1855, the number of known, named, and more or less adequately described species of the Mammalian class was estimated at 2000; it is now 3500. The examples of each species required for students and scientific men, include a male, a female, and the young, where the characters of immaturity, unknown as such, are likely to mislead. These will amount to about 8000 specimens for the illustration of the class as it is now specifically determined and named. In the British Museum, there are about 3000 stuffed specimens exemplifying about 2000 species of the class Mammalia; the giraffe being the largest, and a walrus of 14 feet in length occupying the greatest exhibition-space as a single specimen. The rooms assigned to this class are equivalent to a gallery of 220 feet in length, by 35 feet in breadth, giving a superficial area of 6600 square feet. In this space, as is notorious, the specimens are packed as closely as they can be stored, often three, four, or five deep in the cases; or they crowd the floor like a herd of cattle; or they are attached to the wall, at heights inaccessible to the scientific observer. The arrangement of this proportion of the Mammalian class, in single series, so as to give every requisite facility to the student, would require a gallery of about 450 feet in length, by 45 feet in breadth.

The Mammalian organisation is that under which the individual animal is enabled to attain the greatest bulk ; this striking feature of certain of its members is one of the peculiarities of the class. For purely scientific purposes, size needs only to be accurately measured and recorded ; but a Museum destined to gratify the curiosity of the people, and afford them subjects of rational contemplation, ought to exhibit the maximum of the characters afforded by the dimensions of certain species of a class which is peculiar for the vast bulk of such species. The largest specimen of a whale that can be procured ought, therefore, to be exhibited in a National Museum.

One specimen, at least, out of the number of genera and species of the whale-kind should be prepared as an example of the power of the Creator as manifested by the hugeness of the creature. I have visited Museums in which there were pre-served as many of the smaller Cetacea, stuffed, as would make their collective skins more than equal in extent to that of a Mysticete whale of 90 feet in length, and they had been so prepared as to be free from offensive odour.* The same care and skill

---

* Noting the ideas on this subject of some Hon. Members in the

would, I apprehend, be attended with the same result in a properly-stuffed whale. There is another

---

British Museum Debates of July 22, 1861, I applied to M. Schlegel, the accomplished Director of the National Zoological Museum of the Netherlands, for a return of the exhibited specimens of Cetacea in his galleries, and received the following reply :—

"Leiden, the 27th of July, 1861.

"In our galleries the stuffed specimens of Cetacea, and also the larger skeletons of these, are hanged at the plafond of the building, with the exception of the skulls and bones of whales, which are exposed on platforms or tables, or simply placed upon the floor. The smaller skeletons of Cetacea are placed in the ordinary cases behind glass. The stuffed specimens occupy a gallery apart. The skeletons and skulls take about 200 feet length, and 25 feet breadth of their gallery, English measure. It must be observed, however, that the wall-room of these galleries is on one side occupied by ordinary subjects of Natural History of other kinds.

"LIST OF SPECIMENS of Stuffed Cetacea, Skeletons and large Skulls of Cetacea, exhibited in the Netherlands National Museum at Leiden.

STUFFED SPECIMENS.

1 *Delphinus tursio,*   adult.
1   „   *melas*   „
1   „   *acutus*   „
1   „   *gangeticus,* juv.
1   „   *delphis,*   adult.
1   „   *phocœna*   „
3   „   „   juv.

LARGE SKULLS.

1 *Balœna australis,*   very large.
2   „   *mysticetus,* comparatively small.
3 *Balœnopterœ.*
    Maxilla inferior, os occipitale, vertebræ, &c., of very large specimens of *Balœna mysticetus.*

SKELETONS.

consideration which impels the Naturalist to urge the acquisition, more especially, of a specimen of the Mysticete or Great Whalebone whale, while yet it may be possible to procure one, viz., the fact of the rapidly diminishing numbers of this species, and the probability of its utter extinction at no very remote period.

The Mysticete whale is confined to the seas of the northern hemisphere; the equatorial mass of

SKELETONS.

1 *Balœna australis.*
1 *Balœnoptera physalus.*
1   „   *boops,*   adult.
1   „   *rostrata*   „
2 *Monodon*   „
1 *Hyperoodon*   „
1 *Delphinapterus japonicus.*
3 *Delphinus melas,*   adult and juv.
1   „   *leucas,*   adult.
1   „   *orca*   „
2   „   *tursio*   „
2   „   *obtusus*   „
1   „   *delphis*   „
3 *Delphinus phocœna,*   adult and juv.
2   „   *acutus*   „
2   „   *Heavisidi*   „
2 Dugong.
2 Manati.

" (Signed)   H. SCHLEGEL,

" Director of the Museum."

tepid water is an effectual barrier to its migration southwards. The peculiar character of its food—small floating mollusks—further restricts it to high northern latitudes. The Mysticete produces but one at a birth, and that at an interval, as is .supposed, of about three years, the gestation extending probably over one year. This is the whale which, on account of the quantity and quality of the blubber, and the large size of the "whalebone" substitutes for teeth, has been most sought after by the whale-fishers. Modern inventions, such as the screw-propeller, improved adjustment of firearms to harpoons, explosive shells, and other effective methods of capture and slaughter, have rapidly thinned the numbers of this huge, but timid and almost harmless animal. Latitudes, in which a century ago the whale-ships reckoned upon and usually made full cargoes, are now traversed by them, as being wholly barren of sport, on their way to the extreme northern inlets, at once the discovery and chief practical remunerative results of our late Arctic voyages. Here only is there now the chance of seeing the spouting of the " right whale." The Natural History Museum of the nation which has

invested most capital and received the largest
returns for it in the slaughter of the *Balæna
mysticetus*, might be expected to be that in which,
when the animal itself had utterly passed away,
would be preserved as a subject of wonder, contemplation, and study, to succeeding generations,
the sole and unique taxidermal evidence of this
marvel of creation.

St. Petersburg justly boasts of the stuffed skin
of its unique Mammoth : Madrid was as famous
for its once unique Megatherium. Any capital
in Europe would be eagerly visited by the Naturalist if a single specimen of the extinct Dodo
were preserved in its national museum ; but a dried
head at Oxford, and a dried foot in the British
Museum, are the sole examples of the skin of this
strange bird which now exist. If the wealthiest
maritime and commercial country should not think
it worth while to preserve one specimen, or at
least the skeleton, of the right whale, before the
species becomes mere matter of history, it will be
amenable to the same reflection which is cast upon
the Dutch, who, intent only on killing the dodo to
victual their ships as they touched at their "Prince

Maurice's Island," *en route* to Batavia, finally extirpated the species about A.D. 1660, without caring to preserve a single specimen in their museums for the behoof of natural science. They have since nobly vindicated their claim to rank among the civilised communities caring for higher matters than material wealth, cherishing the sciences, and especially Natural History, by the liberal scale on which the State of the Netherlands has provided for its Zoological collections.

The diversity of structure shown in the different cetaceous genera, *Balæna*, *Balænoptera*, *Physeter*, *Hyperoodon*, &c., are best exemplified in their skeletons; but, by reason of their size, such require for exhibition the resources of a National Museum. Here only can an intelligent public, or the special student of this least known and most difficult branch of mammalogy, expect the means of contemplating the characters and structures of the strangest as well as hugest of animals, —the most seldom seen by reason of their ocean haunts; air-breathers, yet living in water; hot-blooded, though ever surrounded by a rapidly cooling medium; the most closely allied to man's

own class in all the essential parts of their organization; mammalian, but fishes in shape; a comparatively recent development in the series of
life-forms that have succeeded one another on our
planet, and the superseders of the great sea-lizards
in their peculiar office in the ocean police. Space
is the first essential towards fulfilling this exigency, and adding this striking and instructive
feature to a National Museum of Natural History.

The greatest and most philosophical Zoologists
have ever made the Cetaceous order a special
subject of their investigations. ARISTOTLE first
showed that they were not piscine, but viviparous
and mammiferous.* JOHN HUNTER, taking an
equally comprehensive grasp of living nature,
singled out the Cetacea for that series of perse-
vering researches which he, the most scrupulous
and reticent of writers, deemed worthy of commu-
nication to the Royal Society.† BARON CUVIER,
appealing to administrative authorities of congenial

---

* The latter character had not escaped the observation of the maritime people of the Syrian coast. " Even the sea-monsters draw out
the breast: they give suck to their young ones."—*Lamentations,*
iv. 3.

† ' Philosophical Transactions,' vol. lxxvii. 1787.

aims and grasp of mind, exercised his influence to obtain the erection of a special gallery for the osteology of whales in the Jardin des Plantes; and this alone, had he created no other treasure-houses of the wealth of living Nature, would necessitate a visit to Paris by every genuine culti-vator of Zoology. The descriptions of the speci-mens which Cuvier was thus enabled to receive and conveniently arrange, form a very valuable part of his great and original work.* He visited, moreover, every metropolis where a single illustra-tion of the larger whales existed, which he had not been able to procure for Paris. His first journey to London, after the restoration of peace in 1815, included as a prime object, the study of the then unique skull of the Great Greenland Mysticete whale.† This specimen, rendered classical by the description and figures of the founder of Palæon-tology, still remains of necessity in a dark base-ment vault of the British Museum. JOHN HUNTER

---

* 'Recherches sur les Ossemens Fossiles.'
† "La tête de Baleine de Groenland du Muséum Britannique," is described in p. 281, and figured in pl. 226, figs. 9-11, in the 8th volume of the 8vo. edition of the 'Ossemens Fossiles.'

writes :—" As the opportunities of ascertaining the anatomical structure of these large marine animals are generally accidental, I have availed myself as much as possible of all that have occurred; and, anxious to get more extensive information, I engaged a surgeon, at a considerable expense, to make a voyage to Greenland, in one of the ships employed in the whale fishery, and furnished him with such necessaries as I thought might be requisite for examining and preserving the specimens." For HUNTER complains that in the quest of whales, " gain being the primary view, the researches of the Naturalist are only considered as secondary points, if considered at all."

Have we risen in eighty years " of progress " to any higher view ?

Perhaps the oddest of compromises characteristic of an age of accommodation, is that which was proposed to the recent grudgers of exhibition-space, viz., to omit the Cetaceous order in the national series of Natural History, and retain only the smaller showy specimens.

Birds, shells, minerals, are, however, to be seen in any museum ; but the hugest, strangest, rarest specimens of the highest class of animals can only be studied in the galleries of a national one. It is surely unworthy of a great country to profess to have its Museum of Zoology, and to publicly vote, year by year, the sums required for the purchase of specimens; and yet to postpone, year after year, the cost of the simple buildings requisite to render them available.

But it is not the whales alone that make demands for exhibition-space. Some of the landmembers, as well as the sea-members, of the Mammalian class, exemplify the characteristic of hugeness of bulk in the individual.

> " From his mould,
> Behemoth, biggest born of earth, upheaves
> His vastness."

So likewise—

> " Ambiguous between sea and land,
> The river-horse, and scaly crocodile."

> " Not all
> Minims of Nature; some of serpent-kind,
> Wondrous in length and corpulence, involve
> Their snaky folds."

The elephant of Africa and the elephant of Asia are quadrupeds of which the chief peculiarity is their size; and consequently the largest examples should be selected to exemplify the species in a Public Museum. To such specimens the physiological Naturalist would point as exhibiting the maximum of mass that can be supported and moved on dry land by a living animal.

But for this purpose it may be said that one elephant will suffice; and some Naturalists urge that it is only necessary to exhibit the type-form of each genus or family. But they do not tell us what they mean by their "type-form." It is a metaphysical term, which implies that the Creative Force had a guiding pattern for the construction of all the varying or divergent forms in each genus or family. The idea is devoid of proof; and those who are loudest in advocating the restriction of exhibited specimens to "types," have contributed least to lighten the difficulties of the practical curator in making the selection.

Where, for instance, will he find the principle demonstrated and accepted in zoological science which can guide him in his selection of the type-species of the genus *Elephas?*

By one authority he may be told "the type-form is that from which the various species have

diverged," and that it will accordingly be recognised in the species which retains most of the
general structure of the wider group of which the
genus in question is a specialised member. According to this definition, the museum arranger would
see in the coarser divisions of the grinding-teeth
of the African elephant a nearer approach to that
character of perissodactyle dentition as evinced in
the Tapirs, to which the transition from the Elephants is effected by the progressive simplification
of the grinding surface of the teeth in the existing
Mastodons and Dinotheres.

But a second authority will urge that the "type-
form" of any given genus is that which exhibits
the generic characters in their fullest perfection, in
their highest state of efficiency for the part assigned
to the members of the genus in Life's great drama.
Thus the Feline, with the sharpest carnassial tooth
and the most retractile claw, is said to be a more
typical Carnivore than that which, retaining more
of the common ferine characters, is more closely
allied to conterminous Musteline or Viverrine
genera.

On this idea of a type the elephant of Asia

would be selected on account of the greater degree of the peculiar elephantine complexity of teeth which it exhibits, as contrasted with the African species.

Dr. Gray, memorialising, in 1854, for increased space for the zoology in the British Museum, on the ground of the impending destruction of the specimens then, as now, stored in dark basement vaults, uses the word "type specimen," as that which afforded to the founder of the species his description of it.

But, whatever be the meaning of the advocates of the exclusive exhibition of "types," my idea of the scope and appliances of a National Museum of Natural History leads me to view it as the place where the Naturalist ought to find, for example, the readiest means of comparing the most gigantic quadrupeds of Africa and Asia, not only in regard to the structure of their molar teeth, but as to the proportions of height to breadth of body, the shape of the head, the relative size of the ears (which differs extraordinarily in the kinds of elephant), the number of divisions of the hoof in the fore and hind limbs, the shape and proportion of the

tusks, and the degree in which those ivory weapons are developed in the male and female of
each species. The means of affording these subjects of scientific scrutiny and comparison, in
regard to the elephantine genus, involve the
necessity of having a specimen of the full-grown
male and female of both the African and Asiatic
kinds.

The relations of Zoology to other sciences widen
year by year. They have of late become most
interesting in reference to the geographical changes
which the earth has undergone. It becomes a
most exciting pursuit of the Naturalist, when he
gets in quest of evidences that such and such scattered islands are the mountain-tops of a continuous
continent now submerged, and acquires proof that
such submerged continent belonged to a life-period
distinct from that of the uprisen one which lies
nearest to the still unsubmerged mountains of the
former continent. And the investigation of the
specific characters of the few existing elephants
has an important bearing on this philosophical
relation of Zoology to Geology and Geography.
These huge proboscidians roam not only in the

continent of Asia, as in India and Siam, but also in the neighbouring islands of Ceylon and Sumatra.

Evidence has been creeping in of late, indicating or testifying that the continental Asiatic elephant is distinct from the insular one; that the Siamese and Sumatran elephants are not the same species; that the Sumatran agrees specifically with the Cingalese elephant, and that the Siamese agrees with the Indian one. That many species of elephant more decidedly distinct have formerly inhabited India, and have become extinct, there is palæontological proof. And the fossil evidences of such specific variation add to the importance of precise and accurate knowledge of the still existing species. But Naturalists can only settle such questions by comparison of authenticated specimens of the elephants of India, Siam, Sumatra, and Ceylon. Who knows what new light might be thereby thrown on the past and present physical features of South Asia; on the geological dynamics that have caused the present arrangement of dry land and sea; and even on the question of the origin of races and of species!

With these convictions of the important appliances of a National Museum of Natural History,
silence as to its requirements, in the Naturalist
moved thereby, would be culpable. It becomes
his bounden duty to urge the imperative need of
space. It is only the National Museum that can
afford the requisite means and materials for the
study and knowledge of the Proboscidian order.
In addition to the specimens in the gallery of the
stuffed Mammalia, the osteological gallery should
contain the skeletons of the full-grown males of
both the Asiatic and African species ; together
with the skull and characteristic bones of both
male and female, with the dentition, in its different phases of change and succession, from every
known species or variety of elephant, for the purposes of comparison with each other and with
fossil remains. There would be no great difficulty in obtaining the specimens, if only space
were provided for their reception and exhibition.
To pare down the cost of a National Establishment
of Zoology, by excluding the bulky specimens from
the series, is to take away its peculiar and exclusive

function as an instrument in the advancement of natural science.

Next in size to the elephants in the Mammalian class come the Rhinoceroses; like the elephants they are now much reduced in number and restricted in place as compared with former geological periods.

Four kinds of Rhinoceros have been specified in Africa, two of the white kind and two of the black kind; all of these possess two horns. There is also a two-horned rhinoceros in Sumatra, and the rhinoceros of continental Asia is one-horned, as is that of the island of Java and of Borneo. Of these seven or eight species, which, it may be asked, should be selected as the " type-rhinoceros"? Here the same difficulty meets the museum officer as in the case of the elephants. The Asiatic rhinoceroses adhere more closely by their dentition to the general type, the African kinds depart from it by the suppression of the incisive teeth. In this respect, and in their two horns, both of which are single, symmetrical, and medial, the African may be said to be most rhinocerotic; but the Asiatic species is most so in regard to the thickness and peculiar folds of its tuberous integument. How, then, can a single typical example be selected from this series? It is less difficult to choose the species which might be represented by characteristic parts preserved in store—that is to say, one would select one of the two white African kinds and one of the two black African kinds of rhinoceros: but five out of the seven or eight known rhinoceroses ought to be represented by stuffed examples of full-grown males, with the dried heads of the full-grown females to show the sexual character of the horns.

The present state of doubt or ignorance as to the affinities of the continental and insular rhinoceroses of Asia, and their relations to each other, is owing chiefly to the want of any provision for displaying the specimens in the museum of the nation which rules, or mainly influences, the tropical regions inhabited by those singular quadrupeds.*

---

* As illustrative of how closely the progress of accurate knowledge of species depends on the proper extent of exhibition and supply of specimens in a National Museum of Natural History, I

The two-horned rhinoceros of Sumatra offers, of
all living rhinoceroses, the nearest resemblance to

---

subjoin a letter which my former 'Reports' on this subject have called
forth.

" MY DEAR OWEN,   " Calcutta, April 8, 1862.

"It is not very often that I indulge in a letter to you, but I am
very sure that the following discoveries respecting the distribution of
the Asiatic living species of rhinoceros will interest you.

"In your lecture and evidence regarding a National Museum of
Natural History, you refer to *Rh. sondaicus* and *Rh. sumatranus* as if
each was peculiar to the island from which it takes its name ; which is
very far from being the case.

" The true *Rh. indicus* would seem to be confined to the *tarai*
region at the foot of the Himalayas and valley of Adém, *i. e.*, of the
Bráhmaputra River.

"The single-horned rhinoceros of the Rajmáhl hills in Bengal
(where it is now verging on extirpation), of the Bengal Sandarbáns, of
the Indo-Chinese countries and Himalayan peninsula, is the *Rhinoceros
sondaicus* of Java and Borneo.

" Still more common in Burmá is the *Rh. sumatranus*, which
extends at least as high as the latitude of Ramri Island, on the
Yamadouny range which separates the province of Arakan from that
of Pegu (or the valley of the Irawádi). This species exists, also, in the
Malayan peninsula ; but, in the Archipelago, appears to be peculiar to
Sumatra.

" What the particular species might have been that was formerly
hunted by the Mogul Emperor Báber on the banks of the Indus, may
yet perhaps be ascertained by the discovery of bones buried in the
alluvium of that river.

" All three species vary a good deal in the form of skull ; each showing
a broad and a narrow type, with intermediate gradations. The skulls
of *Rhinoceros sondaicus* and *Rhinoceros sumatranus* have never,
hitherto, been adequately represented ; so that the contrast between
them and the skulls of *Rhinoceros indicus* and *Rhinoceros sondaicus*
as figured by Cuvier and De Blainville, is considerably greater than
exists in fair average specimens.

" However

certain fossil kinds found in Europe. When half-grown, this rhinoceros retains a conspicuous coat of short, straight, bristly hair. It is generally known that one, at least, of the extinct European rhinoceroses was covered with hair when full-grown. Such are the interesting facts and relations that make it desirable to preserve and exhibit specimens of the young as well as old individuals of some of the largest species of quadrupeds. What I have said of the Rhinoceros applies to the Elephant. Bishop Heber's first announcement of the young hairy elephant which he met with in the Himalaya

---

"However they may vary in breadth, the skull of *Rh. indicus*, (adult) averages 2 feet in length, measured by calipers from the middle of the occiput to the tips of the united nasals; that of *Rh. sondaicus* does not exceed 1¾ feet.

"I very greatly suspect that the large rhinoceros that was so long in the Zoological Gardens, and which you described (and which was captured in the province of Arakan), was *sondaicus*, and not *indicus*, according to my vivid recollection of the beast, which was not nearly so large as the specimens of true *indicus* which I have seen in this country. The measurement of the skull would at once decide this matter.

"Of course the variation in the form of the skull of the living Asiatic species should preach a caution to Palæontologists. In my forthcoming Memoir on the subject (now with the printer) I have amptly illustrated these variations.

"Yours ever sincerely,

"E. BLYTH."

D

Mountains excited much surprise. The fact had never been exemplified in our Museum. It is a character which, though transitional in the modern elephant, was persistent in the Mammoth or old Northern Europæo-Asiatic elephant.

Let me next refer to the Tapirs. One species is South American, another is Sumatran; they differ much in colour. Which shall be selected for exhibition as the type-tapir? With every desire to carry into practice the economical recommendation to afford exhibition-space only to the type-form of a genus, the Custodian of a National Zoology has to seek for light to direct his choice in this matter. For, say that the American tapir departs least from the general Pachydermal type of colour, the particoat of the Sumatran species might be contended to be the most strictly tapirine peculiarity. But if votes prevailed for the American species, the question would next arise,—Shall we exhibit the tapir of the plains, or the tapir of the mountains? For it is held that the *Tapirus andium* is a distinct species from the *Tapirus americanus*. Obviously, a National Museum of Natural History is the place where Zoologists should find the means of determining or satisfying themselves as to such questions.

It is easy for the Book-naturalist to rule that only the type-form of the genus need be exhibited in a Public Museum. But even were all agreed as to the type-rhinoceros, type-elephant, type-hippopotamus, or type-tapir, the practical Zoologist would testify that it is easier to make the required observations and comparisons of the species on the skins of great Pachyderms when stuffed and set up, than to have to open the stiff folds of a flattened and close-packed hide, and haul the bulky and cumbrous mass about, to bring to view this or that particular character.*

A singular and interesting modification of the Mammalian type for sea-life is that manifested by the Dugongs and Manatees—the mermen and mermaids of mariners. The Manatees are from 15 to 20 feet in length; species haunt the estuaries and

---

* It seems incredible that such an assertion could have been hazarded as the following, by an advocate of the existing state of things. "Students and scientific men greatly prefer to have the specimens for examination in cases occupying but small space in comparison, which admit of their being much more easily handled, compared, and measured."

large rivers of central and intertropical South America, and also those of both the eastern and western sides of tropical Africa. The Dugongs, from 12 to 16 feet in length, have been found in the Red Sea, the Malayan archipelago, and the northern coasts of Australia.

Transmutationists see in these animals special evidences of the course of change, either from imped to quadruped, or from quadruped to imped. Only the powers and resources of a great National Museum of Natural History can afford the philosophical Zoologist the means of comparing and tracing the characters which, on one hypothesis, are the marks of original specific distinction, formulized by the terms *Halicore tabernaculum, H. australis, H. dugong, H. malayanus,* &c.; or which, on another hypothesis, may be the indications of the direction, whether ascensive or retrograde, of the modifications of structure that may be slowly progressing, in long course of time, to lead to what are called new specific forms.

The slow progress of the science of Zoology is partly, indeed greatly, due to the inadequacy of the means of the private individual, or even of an association of scientific Zoologists, to acquire, preserve, display, and arrange the specimens which are the necessary elements of the problems in question. The series of the Sirenia, like the series of elephants, rhinoceroses, tapirs, and cetacea, require as the essential preliminary condition adequate exhibition-space,—such space as a Museum of Natural History professing to be national ought to cover, and can alone be expected to afford.

I will only cite one other group of Mammalia, which bears upon the question of space in a National Museum ; it is that group which, subsisting exclusively on marine animals, offers a most interesting intermediate modification between the fishlike Cetacea and the Carnivorous quadruped. Species of the group to which I allude, viz., the "Pinnigrade Carnivora," Phocidæ, or Seal-tribe, have attracted, since the diminution of the Whale-tribe, almost as much attention from the commercial collector of animal products as the whale itself.

Seals of the kind called "Sea-elephant" (*Cystophora proboscidea*), frequenting remote and desolate islands in the great Southern Ocean, attain the length of 30 feet; and, seen skulling their rapid

course by alternate strokes of the terminal pro-
peller, formed by their connate hind-flippers and
tail, have excited the idea of the "Great Sea-serpent"
by reason of the long track or "wake" left on the
broad billows they have cut through. These large
Seals are slaughtered annually, chiefly for their
blubber, and in such numbers as, with their restricted localities and closer dependence on dry
land than confines the whales, foreshows a speedy
extinction of the species. They are elephantine,
not only by their bulk, but by the proboscidian pro-
longation of their snout, at least in the adult male.

Another huge species of Seal resembles the
elephant in the ivory-like tusks that descend from
the upper jaw. One of these "Walruses" is exhibited, stuffed, in the British Museum ; it occupies
a superficial space of 16 feet by 10 feet ; but larger
examples are recorded to have been shot by voyagers
to those remote Arctic regions to which the Walrus
is now confined. The number of species of Seals
registered in Zoological works is upwards of fifty ;
their diversity of form and structure has led to
their partition into five well-marked families.
Exercising a proper discretion in the selection of

specimens for exhibition, they would range along about 100 feet of the Mammalian Gallery.

In the exercise of this discretion the number of species represented by stuffed specimens, in comparison with those represented by dried skins kept in boxes, becomes less as the species diminish in size. The Rodentia, for example, is the order of Mammalia containing the most numerous as well as the smallest species. A larger proportion of the Mouse family, *e. g.*, and also of the Shrew family from a conterminous order, may be preserved in store, with sufficient display of the modifications of form and character by the exhibited specimens, than would be proper with regard to the groups of few and large species of Mammalia.

Accordingly, the museum curator and arranger finds, that in adapting the selective principle to the limits suited to the aims and functions of a Public Collection, he by no means thereby saves space in the ratio of the numbers of the class that he may reserve in store.

I believe that enough has been adduced from the Mammalian class, to illustrate the conditions on which must be founded calculations as to the extent

of exhibition-space required for the selected specimens from that class. Let me again refer to the
ratio at which the Zoologist's knowledge of the
class has proceeded of late years; viz. from, say
1350 species in 1830, to 2000 in 1850, and to
3500 in 1862. In one order, e. g. *Marsupialia,*
the increase has been from fifty species recorded in
1830, to 350 species in 1860.

We should greatly over-estimate our present
knowledge, were we to rest upon it a conclusion
that there remained but very few more forms of
Mammalia to provide room for in our -museums.
Look, for example, at the recent unexpected aug-
mentation of the species of the Quadrumanous
order, by the researches and acquisitions made by
Dr. Savage and M. du Chaillu, in a limited, but
previously unexplored, tract of tropical Africa,—
species including the largest as well as the most
highly-organised forms of the order that comes
nearest to Man.

One has only to glance at the latest published
maps, and note the proportion of the intertropical
earth yet unexplored, in order to adjust our outlook for accommodation-room, and regulate our

expectations of the new forms and species of animals that may make demands upon the space of a National Museum.

But space is required, not for the mere housing of the Zoological specimen, but for convenient display, accessible to every scrutiny of parts needed by the science; and not merely for such disposition of the individual specimens, but for orderly systematic array of the series. The Galleries should bear relation, in size and form, to the nature or characteristics of the classes respectively occupying them. They should be such as to enable the student or intelligent visitor to comprehend the extent of the class, to trace the kind and order of the variations which have been superinduced upon its common or fundamental characters; to see how the Mammalian type is progressively modified and raised from the form of the fish or lizard to that of man, and to study the gradations by which one order merges into another. Thus, if a gallery of about 400 feet in length, by 45 feet in breadth, be needed for present acquisitions, and some of the most easily-acquired larger desiderata of the class, double that space should be secured for subsequent extension to meet the additions of future years, especially those most needed to complete the Mammalian series.

OSTEOLOGY.—An Osteological collection is as essential to the illustration of the Vertebrated classes as a Conchological one is to that of the Molluscous classes. Nor should the size of an animal be a bar to the obtainment of adequate space for the exhibition of its skeleton in a National Museum of Natural History. In the order Cetacea, the proportion of the skeletons would exceed that of the stuffed specimens, especially in the genera of largest bulk. *Balæna, Balænoptera, Physeter, Hyperoodon,* should be represented by the skeleton of the largest species in each; skulls and separate bones would exemplify the differences on which the other species have been founded. But as the exterior parts of animals are the seat of much more variety than are the internal organs, the proportion of mounted and articulated skeletons in other and more populous orders of animals, required by a National Museum, would be smaller than that of the stuffed and mounted skins. In 1847, the British Museum possessed 1766 Osteological specimens appertaining to 742 species of Mammals. In 1861, it possessed 4255 specimens, viz., 766 skeletons, and 3549 skulls or parts of skeletons. These contnue to be stored in the basement vaults: but a rapid and most useful increase would be made in this department, if even no more space were allotted to it than is shown in the Plan No. 1, Plate I.

BIRDS.—In the National Museum of every country the class of Birds is the best exemplified; in our own it is the only part of the Zoological series which may be said to be exhibited as a whole, or as a class should be seen. The Mollusca are represented by their shells; other classes are exemplified in a fragmentary way. In a badly-lit gallery of 300 feet in length, the vertical or wall space is appropriated to the stuffed birds, and along this wall-case which, through projections in the gallery, is of 900 feet in extent, about 2500 species of birds are exhibited, but in a too crowded manner. This, however, affords a good basis for estimating the extent of an Ornithological gallery for such a display of the known class of Birds as ought to be made according to my view of the scope and applications of a National Museum of Natural History.

Along 900 feet of wall-space, in cabinets 10 feet in height, may be arranged stuffed specimens of say 2,500 species of birds. The British Museum possesses skins of 4,200 species of birds; the number of species now known and described is about 8,300.

Birds are most attractive to the sportsman; the largest proportion of the collection of the foreign traveller usually appertains to the feathered class. The rate of increase of the knowledge of the class may be conceived by the fact that GOULD, in his three years' life among the woods and wilds of Australia, discovered new species, the descriptions of which occupy the chief part of his great work on the birds of that continent in nine folio volumes. Looking forward to the additions to the class of Birds that may be expected from the interior of Africa, from Madagascar, from Borneo, from New Guinea, and from the great proportion of Australia yet unexplored by the Naturalist collector, a vast and rapid increase of singular and interesting forms may be expected, is indeed certain. But on the basis of the class as it is now known, and exercising the same restrictive discretion in the selection of specimens for exhibition, a gallery of

850 feet in length by 40 in width should be assigned to the class *Aves,* or space secured for such ultimate extension of the gallery of Ornithology. For the purposes of science, each species of bird requires to be exemplified by a male and female in mature plumage, and by the young, in one or more stages of immature plumage.

The remark previously made on the osteology of Mammals applies to that of Birds, save that the number of articulated skeletons would bear a smaller proportion to the stuffed specimens. But there are classes of objects required to complete the exhibition of the Natural History characters of Birds, not needed for that of Mammals, viz., the nest and eggs, with some singular preliminary structures in a few species.

The Megapode (*Megapodius tumulus*), the Leipoa (*Leipoa ocellata*), and the Talegalla (*Talegalla lathami*) are Australian birds, which construct mounds of dead leaves and other decaying vegetable refuse, in which they lay, or rather insert, their eggs : leaving them to be hatched by the heat of fermentation. Their eggs are disproportionately large, and the chick acquires from the abundant nutritive

store included in the egg, sufficient organic material
for a grade of development enabling it to fly and
provide for itself as soon as it emerges from its
extraneous nursing-bed.

A pair of Talegallas, or brush turkeys, at the
London Zoological Gardens, occupied themselves
during the spring and summer of 1860 in forming
a large hatching-mound. The females laid and
buried twenty eggs, in a circle, with the small end
downward, and on the 26th of August a young
talegalla emerged from the mound, and ran about
in quest of worms and insects; feeding independently, and without care or concern on the part of
the parents. Towards night the young bird took
flight and settled on a branch of a tree about 6 feet
from the ground. On the 28th of August a second
chick was observed beneath the surface of the
mound, busily occupied cleaning its plumage and
freeing the quill-feathers from the remains of their
formative sheaths : on the morning of the 29th it
emerged, and at once took flight. It then sought
its food, like its predecessor, independently of each
other, and without exciting the notice of either
parent. The growth of these young birds was so

rapid that, after three months, they could scarcely be distinguished from the adult birds.

As incubation must commence from the time of the immersion of the egg in the fermenting mass, the young will emerge in the order of oviposition; that from the twentieth egg escaping as many, or perhaps twice as many, days later than that from the first egg. The advanced state of development and the instinctive independence of action, both accord with these necessary conditions of extraneous or artificial incubation, in the above-cited Australian birds.

But this is not the sole peculiarity observed by Mr. Gould in the bird-population of this singular and exceptional continent. Here are to be found certain conirostral Passerine species which still practice the habit of constructing marriage-bowers, apart from and independent of the later nesting and nursing home.

The satin Bower-bird (*Ptilonorhynchus holosericeus*), of New South Wales, and the pink-necked Bower-bird (*Chlamydera maculata*), of central and northern Australia, are remarkable for their construction, on the ground, of avenues, overarched

by long twigs or grass-stems; the entry and exit of
which avenues or bowers are adorned by pearly
shells, bright-coloured feathers, bleached bones, and
other decorative materials, which are brought in
profusion, often from a considerable distance, by
the male bird; and variously arranged, to attract
the female, as it would seem, by the show of a hand-
some establishment. "It is then most amusing,"
writes the observant Gould, "to witness the antics
of these birds, running in and out of the arched
avenue; the male attitudinizing, setting up his
feathers in the most grotesque manner and dancing
about the nuptial bower."

For receiving and incubating her eggs the female
builds a nest, like that of the magpie, in the con-
cealment of a tree. It is possible, that the odd
propensity of the Magpie, Jackdaw, and some others
of our conirostral Passeres, to which the Australian
Bower-birds are allied, to pilfer glittering objects,
may be the remnant of a similar bower-building,
bower-ornamenting, instinct, which the increase of
human population has scared out of them; the
conditions of our well-peopled, widely-cultivated
island having reduced the daws and magpies to

restrict their constructive propensities to the nests essential for the continuance of the species.

In the Avian gallery of a National Museum a model of a mound should be exhibited, and might be prepared to show the number, size and arrangement of the eggs of the Leipoa or Talegalla; and space should be provided for the exhibition, not only of the stuffed skins, but of the osteology, eggs, nests, hatching-mounds, and artificial bowers, of birds.

REPTILIA.—Of this class, including, as in Cuvier's system, crocodiles, lizards, chelonians, serpents, and batrachians, upwards of 2000 species are now known. In the wall-cases of a gallery of 70 feet in length, about 200 of the less bulky species can be properly exhibited by means of stuffed specimens. But some of the crocodiles attain a length of from 20 to 25 feet. Some of the turtles, also, are of large size. The Python of Africa or Ceylon and the Boa of South America should exemplify their "wondrous length" by the hugest specimens obtainable. The different stages of the metamorphoses of the frog, toad, hyla, pipa, and other leading forms of the batrachians should be exhibited. Those of the common frog should also be illustrated by enlarged

E

wax models for the instruction of the public visitors. A gallery of 250 feet in length by 40 feet in width would be required to illustrate, by selected species, a continuous series of the entire class *Reptilia*.

So likewise, in reference to the class of *Fishes*, the chief stages in the growth of such species as the salmon, which stages are known by the local terms " Parr," " Smoult," " Grilse," &c., should be exhibited. And in several species of this class the male and female are distinguished by well-marked external characters, which should be exhibited. Of 4000 species of fishes in the British Museum, about 1000, represented by 1500 stuffed specimens, occupy the exhibition-space of a gallery 70 feet in length. Others are attached to the wall-surface above the cabinets, where no proper examination or comparison of the specimens can be made. The number of named species of fishes now amounts to upwards of 8000. The largest known fish of the present period is a British species, in so far as specimens have been cast ashore, within the present century, near Hastings * and Brighton,† and at

---

* See ' Philosophical Transactions ' for 1809, p. 177.
† A large specimen was captured here in 1812.

the Western Isles of Scotland; it is the Basking
Shark (*Selache maxima* of Cuvier) which attains a
length of upwards of 30 feet. It is only in a public
national collection that this rare and largest species
of shark could be expected to be exhibited in a
properly prepared state. Amongst the Tænioids,
or ribband-fishes, are species which attain the length
of from 30 to 40 feet. They are among the
numerous and various phenomena, which, caught
sight of while moving on the undulating surface
of the ocean, engender the idea of the " Great Sea-
serpent; " and the ultimate dispersion of such
delusion can hardly be expected, whilst the Sea-
elephants, Basking Sharks, and full-sized *Gymnetri*,
Trachypteri, Bogmari — all of which have been
successively attested, sometimes on oath, by eye-
witnesses, as being the Great Sea-serpent,*—continue to be wanting in the public galleries of the
National Museum of Natural History. In a gallery
of 300 feet in length, by 40 feet in breadth, the
principle of selection would have to be exercised,

---

* See the account of this animal by Dr. Barclay, with the affidavits
of its observers, in the first volume of the 'Wernerian Transactions,'
and the description of one of its vertebræ, now in the Museum of the
College of Surgeons, No. 434, 'Osteol. Catal.,' vol. i. p. 98.

in order to give, in a connected series, all the ordinal and family modifications of the class of *Fishes* including the above-cited illustrations of its large members.

Reptiles and fishes do properly constitute one great natural group of cold-blooded animals. And the beautifully fine series of gradations by which the cold-blooded air-breathers pass through the amphibious series to the water-breathers, are well exemplified in the modifications of the skeleton. Should a special gallery be allotted to the osteology of the *Hæmatocrya*, or reptilian and piscine cold-blooded animals, it ought to have a length of 250 feet, by 40 feet.

INVERTEBRATA.—The reasons which have led, in the allotment of space in most public museums, to the exposition of a larger proportion of the class of Birds than of other Vertebrata have produced the same effect in regard to shells in the Invertebrate series.

The texture, form, colour, and pearly lustre of these skeletons of the mollusca all accord with our ideas of the Beautiful; moreover, their preparation for exhibition demands the smallest amount of skill and time, and they are almost imperishable. The substitution of fresh specimens for those that have faded in colour by long exposure to light is the chief work required to maintain the good condition of a collection of Conchology.

As a basis for calculation of the space that should be alloted in a public museum to the molluscous classes, the readiest is the amount of space so alloted in our own, and its relations to the present known extent of that great primary division of the animal kingdom. After the class of Birds, the display, next in completeness, of a natural group of animals is that which the exhibition of the shells of the Mollusca afford in the British Museum.

The cabinets on the floor of a gallery, which is 278 feet in length by 40 feet in width, are exclusively appropriated to this attractive class of natural objects. Here are arranged and displayed about 100,000 specimens of shells, illustrating about 10,000 species of Mollusca. About 45,000 specimens are stored in drawers or in the vaults.

The well-known private collection of Mr. Hugh Cuming, of No. 80, Gower-street, contains shells, in first-rate condition as to form and colour, of upwards of 16,000 species of Testacea ; and this may be regarded as the present known extent of

the existing species of the shell-bearing mollusca;
but the annual increase of specimens of new species
is great.

There are, however, as I have already said,
classes and orders, as well as genera and species,
of mollusca which have no testaceous covering.
The class *Tunicata*, the orders of *Gymnosomata*, the
Nudibranchs, Inferobranchs, and *Dibranchs,* half of
the *Tectibranchs,* the families, *Limacidæ, Oncidiadæ,*
Firolidæ are all naked mollusca; as are odd genera
here and there in some of the better defended orders.

Of course, in a National Museum of Natural
History, a general view of the whole molluscous
group should be given; and it is more especially
incumbent to afford the public a view of those
modifications of the class which are least likely to
be seen in collections of inferior importance or of
more limited scope. Such modifications include
the exquisite and beautiful forms of some of the
soft and shell-less marine orders of mollusca above
cited. Wax models, coloured after nature, and
specimens preserved in spirits of wine, afford the
means of completing a consistent exposition of the
molluscous sub-kingdom.

As shells, moreover, are but the exo-skeletons or

protective exuviæ of the animal, and as in all mollusks, with very few exceptions, more or less of the
soft parts are protruded in the living animal, and
a large and characteristic part of the animal is so
exposed in the locomotive shell-bearers, the art of
the modeller has been put in requisition in all
public museums to give an idea of the true outward form and character of the mollusk, though it
be usually restricted to one species of a genus or
family. The specimens of mollusks, in which the
soft parts are preserved in some clear antiseptic
fluid, are more numerous; their anatomy, as the
indispensable guide to a knowledge of the nature
and affinities of the Mollusca, has necessitated collections of the entire animals almost co-equal with
those of their mere shells.

I estimate, therefore, supposing a gallery of
not more than 300 feet by 50 feet in breadth
be assigned to all the classes of the molluscous
sub-kingdom, that the addition of the vertical or
wall space (given in such galleries to the class
of Birds in the British Museum) would only suffice
for an equalised and consistent exhibition of all
the chief modifications of the sub-kingdom as it
is at present known; the principle of selection

governing the amount displayed to the same degree in which it has governed the display of the proportion of shells now shown in the table-cases or floor-cabinets of a gallery of 278 feet in length.

Some years ago a Committee of the British Association memorialised the Trustees of the British Museum on the subject of the combination of the fossil with the recent shells.

It is assumed that, practically, the series of stuffed mammals will be kept apart from the fossil remains of that class in all collections of Natural History; and the like with respect to the stuffed birds, reptiles, and fishes—in other words, that the department of Zoology will continue to be distinct from the department of Geology with Palæontology. But adherence to the strict letter of such practical or administrative arrangements would be neither necessary nor wise. Fossil shells readily afford duplicates. The genetic and truly philosophical series, according to geological epochs, may be kept quite complete in the Geological Department; and yet examples of an *Orthoceras*, a *Lituite*, a *Goniatite*, a *Ceratite*, an *Ammonite*, a *Hamite*, a *Baculite*, a *Turrilite*, and a *Belemnite*, might be spared to be most instructively associated with that poor remnant of the order of siphonated-chambered shells which is exemplified by the genus *Nautilus* in the collection of existing Mollusca. Adopting the association of fossil with recent shells, on the same principle, and on the same limited proportion, *i. e.* in order to indicate how the wide gaps in the recent series of molluscous forms had once been filled up, the practical arranger and curator of a National Museum must expand his estimate of space for the Zoological Department accordingly.

ARTICULATA.—Of insects 150,000 species are now known, which is about more than double the number of all the other articulate classes combined —viz. *Arachnida, Myriapoda, Crustacea, Cirripedia, Epizoa, Annelida,* and *Entozoa.* The instances of species remarkable for bulk—such as the great Japanese crab, of 12 feet or more from the tip of one claw to that of the opposite—are rare and exceptional; but smallness of size is met by multiplicity of species. Two entire classes, *Annelida* and *Entozoa,* the sub-class *Epizoa,* some *Cirripedes* and other *Crustacea,* most *Arachnida,* and some insects, especially the larvæ, require to be preserved in antiseptic fluid, or to be represented by wax models.

The exhibitions of the articulate province of
animál life are by no means on a scale proportionate to that of the molluscous and vertebrate
plans of structure. Consequently a very general
ignorance prevails as to even the common forms of
our own indigenous, especially littoral, species of
this province. Almost every year I receive either
specimens or attempts at descriptions of what is
supposed to be an unknown rarity, as indeed it is
the sole British animal whose vesture rivals that
of the tropical humming-bird in the brilliance and
variety of its iridescent metallic colours. It is a
member of the class called *Annelida* or *Annulata*;
it is known to the fishermen of some parts of our
coast as the " Sea-mouse ; " it is the *Aphrodite
aculeata* of Linnæus. If it deserve a name bor-
rowed from the synonyms of the Goddess of Love
by the beauty of its coat, it resembles rather the
God of War in the variety of barbed, spear-like
weapons with which it is provided ; and the stronger
bristles afford interesting objects for the microscope.
Space ought to be provided for the public display,
in well-arranged series, of the various genera and
species of *Aphrodites, Nereids,* tubeworms with their

richly-ornamented crown of tentacles, sea-centi-
pedes and other sea-worms, earth-worms, leeches,
Gordii or hair-worms, and other members of the
Annelidous class. The medical man ought to have
the comfortable conviction that in the National
Museum of Natural History he will find in the
Annulose gallery a well arranged and displayed
series of the *Entozoa*, with which he might be able
to compare the intestinal or other internal parasites
for the relief of the symptoms of which his skill had
been required by a suffering patient. In estimating, therefore, that a gallery of 250 feet in length
by 40 feet in breadth should be provided for the
various classes of the *Articulata*, the principle of
selection must guide the arranger, on a more restricted scale than in the preceding classes, in avail-
ing himself of such extent of space for systemati-
cally arranged and exhibited specimens.

RADIATA.—In assigning also a gallery of 250
feet in length by 40 feet in breadth to the Radiate
province, I have had regard to the size of most of
the species as compared with that of the majority
of the articulate animals, especially insects. Some
Medusæ are 2 feet across; a specimen of the brain-

stone coral (*Meandrina cerebriformis*) may exceed a yard in diameter, and be nearly a yard in height. And of course the finest and largest specimens should be sought out for exhibition in a National Museum.

Of the radiated class of animals, most of the class *Echinodermata* can be represented by their prepared and dried external crusts; the beautiful orders of the Sea-urchins (*Echinidæ*) and Star-fishes (*Asteriadæ*) are so exhibited. But the Holothuriadæ and other soft vermiform kinds of the class need to be preserved in liquor, or represented by coloured models. Indeed, to give an adequate form of the Urchin or Star-fish as it is seen living and moving in its native element, with its hundreds of tubular suckers extended from the ambulacral openings, a wax model of a specimen of at least one of each genus should be added to the series of the Shells.

The numerous, various, and always beautiful kinds of calcareous coral, madrepores, gorgoniæ, fan-corals, &c., ought all to be exhibited. The rather backward state of this branch of Zoology is in great part owing to the non-provision of due exhibition-space in public museums for well-named specimens of this attractive class. Of the class *Bryozoa* the majority of specimens in the British Museum, described in the Catalogue published in 1852, with numerous subsequent additions, are preserved in spirits and stored in the basement vaults awaiting the acquisition of exhibition-space. In the same state is the collection of the entire class of *Acalephæ* (blubber, jelly-fish, medusæ) preserved in spirits.

Whenever we may have command of a gallery of adequate extent for the *Zoophyta*, the zoological curator would take the requisite steps to obtain and exhibit—especially if the Public Collection were that of a maritime and commercial people—illustrations of the growth and structure of coral reefs by means of adequate masses of those beautiful structures, so as to exemplify the law of growth and order of superposition of the different kinds of coral-forming zoophytes which build up the lovely atolls and the dangerous fringing-reefs in various parts of the tropical and warmer temperate latitudes.

PHYSICAL ETHNOLOGY.—No collection of Zoology can be regarded as complete that does not contain illustrations of the physical or natural history characters of the human kind (*Bimanous* order; *Archencephalous* sub-class). Such illustrations are afforded by casts of the entire body, and of characteristic parts, as of the head or of the face, of the hand and the foot; also by the bony framework, and especially the skull, and by the brain preserved in spirits, showing its characteristic size and distinctive structures. Casts of the entire body, for the aims of the Ethnologist, should be coloured after life, as in the examples of the Ethnological Department in the Crystal Palace at Sydenham. The skeletons of every variety ought to be arranged side by side for facility of comparison. This facility is the greater when the separate bones are also similarly arranged in series. The attention of philosophical travellers has at length been directed to this exigency in Natural History; and, judging from the number of instructive specimens brought from South America by Sir Robert Schomburgk, and from India by the brothers Schlagintweit, I have estimated that the Gallery of Physical Ethnology should be 150 feet in length by 50 feet in width in a National Museum of Natural History.

BRITISH ZOOLOGY.—Such a museum, moreover,

requires a special department for the illustration of the Natural History of the people's own country. In the British Museum there is assigned to the Zoology of the British Islands a gallery of 84 feet in length, by 24½ feet in width. It is well filled with the specimens which can be exhibited in a dried state. I shall afterwards refer to the location of the department illustrating the Fauna of a country, in the second plan of a National Museum of Natural History (Plate II.).

BOTANY.—A Museum of Natural History must include for its consistent completeness a Department of Botany; that is to say, a gallery for a collection illustrative of the characters, structures, and economy of the vegetable kingdom as it now exists. The Department of Palæontology includes the series of the fossil remains of extinct plants. Our knowledge of those remains may be not less limited than that of the least known classes of fossil animals. Each new phytolite that is found may call for comparisons with the most extensive series of existing plants for the elucidation of its nature and affinities. The collections, therefore, of prepared and preserved plants and their parts ought to be in the same museum as the collections of fossil plants. My estimate of the requisite space, founded on that now allotted to the Botanical Department in the British Museum, and on the space allotted to the Museum at Kew, led to the proposition of a gallery of 250 feet long, by 50 feet wide, for the recent plants. I am assured by competent and practical Botanists that I have not erred on the side of excess of space in this estimate.

GEOLOGY AND PALÆONTOLOGY.—The progress of these sciences has created a new demand for space in Natural History Museums, which demand has been more or less liberally responded to in the museum of every metropolis in Europe; first and most in Paris, honoured as being the seat and scene of the immortal labours of the founder of the science of fossil remains. The impulse given by Baron Cuvier to their acquisition has been followed by the formation of most extensive and instructive collections of the extinct members of the animal and vegetable kingdoms. Of extinct animals only there are at the present date not less than 6000 genera named, and more or less clearly defined. In many of these genera the number of species already named exceeds—save in the case of a few of the larger genera of insects, i. e. *Curculio, Linn.*—that of the known species of any genus of existing animals. Of this, the genus *Ammonites* is an instance; whilst many larger groups, e. g., *Brachiopoda, Pisces ganoidei*, &c., are almost exclusively represented by extinct species. In fact, of the animals that have calcareous or fossilisable parts in their organization, the number of known extinct species now far exceeds that of the known existing forms; and the knowledge of such fossil species proceeds at a more rapid rate. In 1860 the British Museum possessed 120,000 registered specimens of fossils; of which as many were exhibited as could be packed in cases along an extent of wall of 899 feet linear, and in 34 table-cases, occupying a small proportion of the floor of the Gallery of Mineralogy. At the present date the collection consists of 153,000 specimens of fossil remains; the number of specimens exhibited to the public visitor being about 50,000.

Were space in a Public Museum estimated upon a prospective view of the extent to be provided for the reception and systematic display of

F

annual accessions, a larger proportion must be allotted to the galleries of Palæontology than to those of Zoology. And this is in strict accordance with the nature of things. Existing animals represent the phase of life of the present world, or of the modern epoch in our earth's history. Extinct animals belong to at least a score of equivalent epochs, or to as many former worlds.

If, therefore, for a consistently proportional exhibition of all the classes of existing species of animals, according to the principles of selection adopted for those classes that are best displayed in our own and other National Museums, a series of galleries, collectively of the extent of 3000 feet in length, by 40 feet in breadth, be required, it will be understood that the selective principle must be more restrictively applied to the exhibition of the fossil illustrations of the extinct Zoology in a gallery of 850 feet in length, by 45 feet in width. In such a gallery, however, I have calculated that a series of rock specimens, illustrative of the structure of the earth's crust, with a corresponding selection of characteristic fossils, arranged stratigraphically, might be exhibited, together with the general Palæontological series, arranged according to genus, order, and class.

No triumph of science has appeared more marvellous to the intelligent mind than the reconstruction of a form of life that has passed away long ages ago, and the representation to the visual sense of such animal by its framework, so as to leave little to the imagination in realising a complete idea of the once living figure of the extinct beast. In the British Museum, the North American *Mastodon*, the South American *Megatherium*, the Irish Giant-stag (*Megaceros*), the New Zealand Giant-bird (*Dinornis*), are thus exhibited. In the Museum of the Royal College of Surgeons, the great Ground Sloth (*Mylodon*) and the giant Armadillo (*Glyptodon*) are similarly displayed. At St. Petersburg, the huge, hairy, and woolly Mammoth and tichorrhine Rhinoceros—both colossal beasts, once inhabitants of Britain—may be seen; even the thick hide of the first, and part of that of the second, are there preserved.

Paris, Berlin, Turin, and some other Continental museums, can boast also of their more or less complete and unique skeletons of other extinct species of gigantic quadrupeds.

It is the common experience of officers of National Museums that no specimens of Natural History so much excite the interest and wonder of the public, so sensibly gratify their curiosity, are the subjects of such prolonged and profound contemplation, as these reconstructed skeletons of large extinct animals.

Happily for the fulfilment of this, perhaps, by some administrators, viewed as the primary aim of a Public Museum, it can be gratified at comparatively small cost. A fossil bone and a coloured plaster-cast of it are not distinguishable at first sight, scarcely by sight at all. The artificial junction of a series of casts of the bones of an unique fossil skeleton, produces a result equivalent, for all the purposes of public exhibition, to the articulated skeleton itself. Thus every capital in Europe, the Public Museum of each civilised community, may show to the people the proportion of the creatures of former worlds that science has so restored. Requisite space for exhibition being provided, reciprocal interchange of casts would soon furnish such museums with the co-adjusted frameworks of the most remarkable extinct animals that have hitherto been reconstructed.

The following has been the ratio of increase in the collections of Palæontology in the British Museum, chiefly through the grants by Parliament for purchases, during the last five years :—

Date.					No. of additions.
1857	..	..	..	..	..	9,880
1858	..	..	..	..	..	7,300
1859	..	..	..	..	..	3,600
1860	..	..	..	..	..	10,000
1861	..	..	..	..	..	4,432

MINERALOGY.—Minerals, like Birds and Shells, are amongst the most beautiful classes of natural objects. At least those pure, and for the most part crystallised bodies, representing the elementary composition of rocks, have always attracted the eye and excited the desire of possession, by the combined qualities of form, colour, lustre, and imperishable density, and by the conventional value attached to extreme rarity ; and, perhaps, it may be because they are objects of beauty and price, that the Mineral series rivals in the proportion which is exhibited, and in its connected arrangement, the other two exhibited classes of natural objects, Birds and Shells, in our British Museum. The present Gallery of Mineralogy is 270 feet in length, by 36 feet in width; and in 1857, there were exhibited in the cabinets arranged in that space, about 15,000 specimens. In 1858, 1000 specimens were added; and in 1859, the famous Allan-Greg Collection, of 9000 specimens, was purchased, and is now in course of incorporation; these, with additions from other sources, have already raised the collection to upwards of 50,000 registered specimens of choice Minerals.

From known collections of repute that may be, but are not yet, acquired, an equal ratio of increase is likely to continue.

A public collection of Mineralogy, to be complete in its application, should possess multiples of the same species, in sufficient numbers and of a sufficient number of species to be arranged in distinct series, having different aims. 1st. The *Classificatory* series; including the main bulk of the collection, exemplifying the natural arrangement of Minerals, according to the system adopted by the curator of the department, most probably on the combined bases of chemical composition and geometrical forms. 2nd. The *Geometrical* series; selected to illustrate the varieties of crystalline forms exhibited by each mineral. With which should be associated a series of models, representing these varieties of form in symmetrical proportions, on a scale dispensing with the need of a lens, easily addressing the eye, and requiring from a student of such series only the use of the hand-goniometer. 3rd. The *Elementary*, or elementarily instructive series. The elementary labours of the student should be further facilitated by the series illustrative of the various physical characters whereby the Mineralogist discriminates between different minerals and determines the external features, such, *e. g.*, as the degrees of lustre, the varieties of cleavage and of colour. 4th. The *Technological* series. Minerals are, of all substances, the most determinate and durable in regard to colour; of which they exhibit the most complete series of gradations and varieties. Of this quality Werner took advantage in proposing his system and classification of colours, which in a Public

Museum ought to have a special series for its illustration, with Werner's nomenclature and its synonyms in different languages; so that manufacturers and others might see and know precisely the colour and shade signified by such terms. Minerals, of all substances, best lend themselves to this economic application. 5th. The *Teratological* series. In those original papers of JOHN HUNTER that have recently been published,* there is a chapter " On Monsters," divided into those of the Animal, Vegetable, and Mineral Kingdoms. The illustrations of the latter by the great Physiologist are those " defective formations" or " disposition of one crystal to form upon another," now included in the section of the science termed the " Teratology of Minerals." The series illustrative of this subject should include the specimens that have assumed deceptive forms by undue enlargement of particular planes, by the development of a twin-structure, or through other complex departures from normal form, requiring the goniometer and geometrical aptitude to fathom the right solution.

---

* ' Essays and Observations on Natural History,' &c., by John Hunter, F.R.S., 8vo., 1861.

6th. The *Pseudomorphic* series. Perhaps the most interesting of all these ancillary series to the philosophical Mineralogist, would be that which should exhibit the various kinds of mineral " Pseudomorphs." Here would be exemplified the transitional capacities of the most obdurate and seemingly unalterable and imperishable of substances; silex replacing or replaced by other and very different stones; the hardest gems becoming degraded by the substitution of matter less dense and durable than their own; the least stable minerals, mimicking by their external forms those that are the hardest and most imperishable, &c.

Thus, multiples of the same species of Mineral would do duty in a National Collection, in the Classificatory, the Geometrical, the Elementarily-instructive, the Technological, the Teratological, and Pseudomorphic series. Moreover, as in the Zoological department, a small room is proposed to be so arranged as to make the microscopical inspection of animalcular forms easy to the public, so a corresponding arrangement in the Mineral department might give as easy an opportunity for observing the exquisitely beautiful phenomena of polarised light; so far, at least, as they are applicable to the discrimination of minerals.

It may be a question whether "artificial minerals" should have a place in this department: by which I mean the crystallized products of the laboratory, and of processes carried on in Metallurgy and other arts. But of another class of compound mineral bodies there would be no such question. In the Mineral department of all public museums are located those most interesting and mysterious bodies, called "Aërolites" or "Meteorites"—the mineral substances which our earth receives from time to time from extra-telluric space. Already, through the energy of Professor Storey-Maskelyne, the series of these Skystones in the British Museum now rivals that extensive and formerly unique collection at Vienna; where I learn that they are enlarging the exhibition-space of their famous Mineralogical gallery to twice its former extent.

With the Gallery of Mineralogy must be associated a room with a special light applicable to the goniometrical observations; a laboratory for investigating the chemical composition of minerals; and a lavatory, with the lapidaries' tools and machinery.

Basing an estimate, therefore, on the proportion of Mineral species which can be exhibited, with the single aim of a classified series, on the ground-floor of a gallery of 270 feet in length, by 40 feet in breadth, and on the known ratio of increase,—admitting also the necessity or advantage of the several distinct series required for the full fruition or completed applications of a National Collection of Mineralogy,—I estimate that it should have appropriated to it a gallery of 550 feet in length, by 45 feet in breadth.

On the foregoing grounds of calculation, with the forecast of probable accessions during thirty years, for which provision of space should be made, it appears that a Museum of Natural History, embracing all the objects of the science from Man to the Mineral, and affording to each class exhibition-room for the specimens selected to show the extent of the class and the kind and degree of variations therein, would require a building, which, if of two storeys. would ultimately cover about five acres of ground.

Such was the main result of the ' Estimate with a

Plan,' printed in a Return ordered by the Honourable the House of Commons, on the 16th March,
1859. The calculations on which it was based were
entered upon at the time when the questions of the
need of space for a National Library and Reading
Room, a National Collection of Antiquities, a
National Gallery of Fine Art, a National Museum
of Technology or applied Art, were being agitated,
and, in various ways, pressed upon the notice of
administrative authorities and of the public.

There appeared to be a general ignorance or
misapprehension of the real state and necessities of
the case, as regards Natural History ; and I felt that
I should be wanting in my obligations to the cause
of the science to which my life has been devoted,
if I omitted, at this crisis, to set forth the statistics
or constants on which an adequate conception of
the requirements of space for a National Collection
of Natural History could be formed.

Nor let any one deceive himself, or try to deceive
others, as to this cardinal point in the fortunes of
Natural History. The conditions of increase, which
have operated in the last ten years, will continue
to influence the next twenty or thirty years ; and my

successors may then, with their increased expanse
of knowledge and opportunities, smile at the modera-
tion or inadequacy of the present outlook. And
this at least may be averred, that the considerations
which have led Parliaments to grant, and Trustees
to allot, the annual sums for the purchase of Natural
History specimens, will be as urgent for those years
to come. Therefore, it is, that I have continually
pressed, on the grounds of economy of both time
and money, for the acquisition of at least five acres,
for the extension, as from time to time required, of
the Natural History galleries: although, for the
arrangement and display of a due proportion of
the several classes of specimens, now collected, a
building upon two acres may suffice. It is on this
scale, therefore, that I have laid down the plans of
a series of galleries, the most simple and suitable
which my experience can suggest for the purpose,
in Plates I. and II.

I am not of those who ignore Architecture as one
of the Fine Arts, and think it comes by Nature;
or who regard its professors as obstructive to the
acquisition of a useful purposive public edifice. I
feel too great pleasure in the contemplation of the

Dome of WREN, and the Tower of BARRY, and
suffer too much pain from the structural deformities
that disfigure some fine sites and fair localities in
this metropolis, to have any sympathy with the
detractors of the genuine masters of Architectural
Art. To whom, therefore, I offer an apology for
having omitted to seek the aid of any of them, in
the attempt to address the eye, by the plans of the
galleries most suitable for Natural History collections, as adapted to the form of the ground which
might be obtained at Bloomsbury (Plate I., fig. 1),
and which it was proposed to purchase at Kensington (Plate II., fig. 1), for the National Museum
of Natural History.

If sufficient space in any suitable locality could
have been had, it would have been preferable to
place in sky-lit galleries all the classes of natural
objects ; but economy appears to compel the choice
of a two-storied edifice for a Metropolitan Museum,
in which, therefore, only half the number of galleries can be lit in the manner most advantageous
for the display of their contents.

The Collections which would be seen with least
disadvantage by side-lights are :—

Mammalia (beasts, whales).

Reptilia (lizards, serpents, toads, &c.).

Pisces (fishes).

Osteology (skeletons, skulls, and other bones, and teeth of vertebrate animals).

Ethnology (skeletons, skulls, models, casts, of the varieties of mankind).

Geology (rock-series and specimens).

Palæontology (fossil remains of animals and plants).

The Collections which ought to be displayed in sky-lit galleries are :—

Aves (birds, with their eggs, nests, &c.).

Mollusca (cuttle-fishes, slugs, and other soft or naked mollusks; shells of the testaceous mollusks, or shell-fish).

Articulata (insects; spider-class; centipedes; crab and lobster kinds, or crustaceans; barnacle-class— sea- and earth- worms, leeches, &c. (*Annelides*); intestinal worms, &c. (*Entozoa*).

Radiata (sea-urchin and star-fish tribes (*Echinoderms*); sea-nettles (*Acalephæ*); sea-anemonies; madrepores, corals, corallines, sponges).

Plantæ (botanical herbarium; structural series; sections of woods; seeds, fruits, &c.).

Mineralia.

For further economy of space, an oblong plot of ground should be chosen of about 1000 feet in length, and of not less than 160 feet in depth, allowing a series of short galleries to extend at right angles to the long or principal galleries, as in Plate I., fig. 1, and Plate II. The lower or side-lit galleries, Plate II., fig. 3, L., should be 30 feet in height, the upper or sky-lit galleries, *ib.* v., 25 feet.

The side-windows of the ground-galleries should perforate, at proper intervals, the upper 10 feet of the walls, leaving 20 feet of wall-space beneath ; and, by the above arrangement, the light would be admitted both by the front wall and by the parts of the back-wall looking upon the interspaces of the short galleries. These, on the ground-floor, would be lit at the same height, on each side : and their breadth being 40 feet, it is assumed that an interspace of 20 feet would suffice for the required light, as in Plate II., fig. 3, section A, A.

The principal galleries are needed for the Mammalia, for the Geology and Palæontology, for the Birds, and for the Minerals.

In the ground-floor plan, Plate I. fig. 1, by including the terminal short gallery at each end, with the front gallery along each wing of the building, there are acquired, in the part of the Museum to be first constructed, two galleries ; each of 325 feet in

length, having 45 feet width in the longitudinal
extent of 210 feet, and 40 feet width in the transverse portion of 115 feet in length. One of these
galleries should be appropriated to the present
collection of the Mammalian class : the other to
the present collections of Palæontology. Between
the terminal or transverse parts of the principal
galleries and the central part of the building there
is space for six galleries, each 115 feet in length,
by 40 feet in width, with intervals for light of 20
feet. One of them is appropriated to Ethnology ;
a second to *Reptilia*, including the Osteology of
that class ; a third to *Pisces*, and their Osteology ;
leaving two galleries for the present collections of
Mammalian Osteology, in either of which might be
exhibited the skeleton of the whale (*Physalus anti-
quorum*), 75 feet in length, and the skull of the
great Northern whale (*Balæna mysticetus*), both of
which are now in the vaults of the British Museum.
In the upper floor of the building, on this plan, are
acquired sky-lit galleries in the same number and
of the same extent as those below. One of the
long galleries should be appropriated to the present
collection of Ornithology, the other to the Mine-

ralogy. Two of the short galleries would receive our present collections of *Mollusca*; one gallery should be assigned to the *Articulata*; one to the *Radiata*; one to the Oology, and to the nests and nidamental structures of Birds, Insects, Mollusks, &c.; leaving one for Botany, as indicated in the Upper-floor Plan of the Museum, of the second locality, fig. 2, Plate II.

In assigning the elevation of these stories, I assume that no narrow gallery, fenced by railing, will be affixed to the wall : it is neither desirable nor safe in a Public Museum frequented on certain days by thousands of visitors of all ages.

Such is the idea of a building adapted to afford space for an equal proportionate display of well-arranged and well-exhibited select specimens of every class of Natural History objects now possessed by the British Museum.

In exemplifying this idea, I have taken such an oblong space of ground as would be afforded in two localities—Bloomsbury and Kensington. On the supposition that the land surrounding the British Museum, bounded by the streets named in the Plan, fig. 1, Pl. I., were purchased by Parlia-

ment, I find the requisite plot of about 800 feet by
160 feet, to the west of the building—on two-thirds
of which plot might be erected the building needed
for the present collections of Natural History; with
the power of adding gallery to gallery, of the
same simple and inexpensive kind, on the remaining area of 285 feet, by 160 feet; and also of
continuing such structures, when requisite, at a
future period, on the land to the north of the
Museum, if reserved for that purpose. In the
building proposed to be erected, in the first instance,
there would be space also, in the centre, for a
Lecture-room of about 60 feet diameter. The
offices for the several Keepers and their Assistants,
with the Laboratory and Goniometrical-room for
Mineralogy, might be obtained at the ends of some
of the shorter galleries, as indicated in figs. 1 and 2,
in Plate I. and Plate II. The Residences and the
Library already exist in this locality. Should it
ever be determined to place the Natural History
Museum under a distinct administration, or should
it be deemed advisable to have a separate or direct
entry to it for the public, it needs only to turn the
plan of the building (Pl. I., fig. 1) to the right-

about; and make the entry from Charlotte Street. The long galleries would then be next that street, the ends of the short galleries next the present Museum, and the long side-lit galleries on the ground-floor would receive their principal light from the street, instead of from the narrower space next the Museum. Under any circumstances, this might be the preferable arrangement.

In the second locality, at Kensington, it has been proposed to purchase for the National Natural History Collections the oblong plot of ground, 850 feet by 190 feet, on which might be erected the building of two stories according to the plans, figs. 1, 2, and 3, Pl. II.; giving rather more accommodation, and subserving the display of a greater proportion of specimens than in fig. 1, Pl. I. For example, each of the long galleries in figs. 1 and 2, Pl. II., is 370 feet in length, and each of the short galleries 150 feet in length (offices included) : the breadth of the galleries is the same as in Plan, fig. 1, Pl. I. The space in this locality also allows the erection in the primary building of a large central apartment with a domed roof, susceptible of arrangements, analogous to those adopted in the

Museum of Economic Geology, Jermyn Street, for throwing light from the circular Museum above (fig. 2, A) into the Lecture-theatre below; which essential adjunct to a Museum of Natural History might here be gained of a diameter of 80 feet (fig. 1, 'Theatre.') The well-lit circular-domed Museum above, of 100 feet diameter, would serve for the reception of an Elementary Collection illustrating the characters of the Provinces, Classes, Orders, and Genera of the Animal Kingdom, and also for the Collections of the Natural History of the British Isles. This part of the building would be of an insulated, fire-proof character, with gas laid on for lighting both the Theatre and the Elementary Museum, whereby both might be made subservient and available for public admission after the hours of daylight.

An Exhibition-room of a circular form is that which admits of the most effective and economic supervision; and the series of specimens there proposed to be displayed are of a nature that would be most profitably shown to, and studied by, the wage-classes after the hours of work. Specimens exemplifying the most striking cha-

racters of the several larger natural groups are
seldom amongst the rarities, usually amongst the
duplicates, of Natural History; and such specimens,
accompanied by fully instructive labels, and where
needed, diagrams, would best fulfil the wish of the
Legislature as expressed by the 'Report of the Select
Committee on Public Institutions' of 1860, and
reiterated by Honourable Members in successive
annual debates on the British Museum.

The arrangement and application of the galleries
of the upper-floor being the same in both localities,
only one plan was needed for that storey, and I have
drawn it as adapted to the building at Kensington,
fig. 2, Plate II. This might allow a certain additional length to the galleries for the Birds and
Minerals, if the space above the Entry-hall were
not appropriated for a Board-room of Trustees,
whether inspective or administrative.

The mode of lighting the upper-floor galleries
is by the introduction of sky-lights at the angle
between the wall and roof (fig. 3, U), as in the
Hunterian Museum in Lincoln's Inn Fields, and the
Geological Museum in Jermyn Street. It would,
also, be practicable to utilise the interspaces of

20 feet diameter, between the short galleries, by arching them over with a roof of glass, spanning from below the side-windows of the ground-floor (L), as indicated in the longitudinal section, fig. 3. In the locality at Kensington, it was proposed also to purchase the contiguous plot of ground, 144 feet by 120 feet, which might be appropriated for the Library and official Residences, should Administrative power finally determine on that locality.

In regard to site, Natural History Science is chiefly concerned in its sufficiency; Administrative Science in its locality. Naturalists desire to have space for convenient arrangement and display of every class of the objects of their study in their due proportion. Administrators have regard to the convenience and easy access of the public; perhaps to the cheapness of the ground.

Not that the cultivators of Natural History have limited their views exclusively to their own special interests in their endeavours to obtain adequate space for the collections. They fulfilled their duty at the earliest opportunity in submitting to the Government, in 1858, in an extensively-signed memorial, the reasons that led them to view a particular locality as offering the greatest convenience to special students as well as to general visitors of the galleries of the National Collection.

The locality of the Museum of Natural History has been made a party-question, and my name has been cited, both in and out of the House, as an advocate for or against this or that particular position. I never gave any grounds for such averments, having always considered sufficiency of space as paramount to any consideration of particular metropolitan position. But I have opposed in every legitimate way those who would sacrifice the advantages of space, even of the proportion most pressingly called for, to a continuance of the collections in the present building.

The following appear to me to be the order of value of the considerations bearing upon this question. First :—Sufficiency of space. Second :— The most convenient access to the greatest numbers. Third :—Contiguity to the National Library. Fourth : — Administrative constitution and convenience. Fifth :—Cost of site. Sixth :—Light, and clean air.

To begin with the last. It is placed there, because the Museum of a coal-burning metropolis, which does not avail itself of the means devised to consume the whole of the combustible, cannot do its duty " out of fire." If erected at the boundary of a suburb sending up the least smoke, access becomes difficult to the majority; and a change of wind may any day negative the advantage sought for at the expense of more important considerations. The inevitable conditions of a London Exhibition of Natural History call for the best arrangements for the admission of light, and stimulate the ingenuity of inventors of the doors, lids, and hinges that most effectually exclude soot.

Fifth. The space needed for the Museum may be cheaper at the periphery than at the centre of the metropolis; but the difference of cost will only affect the estimates for the year, and, perhaps, not at all affect the tax-payers of such year; whereas, succeeding generations are concerned in the second consideration. Only when the arguments for the convenience of different localities are equally balanced should the question of cost enter the scale :

the cheapness of an inferior article is usually a delusion, and repented of in the long run.

Fourth. Partiality to a particular administrative machinery weighs with some, in reference to the locality of the National Establishment of Natural History. And, if the National Establishments of Astronomy, of Practical Geology and Mining, of Horticultural Botany, were worked and controled by a simliar machinery to that affecting Natural History, there might seem to be reason in the objections to severance of a part of the British Museum on account of its being calculated to withdraw such part from the operation of the administrative machinery originally adapted to the totality of the British Museum.

The Legislature may, however, prefer to work the National Museum of Natural History after the analogy of the Royal Observatory, the Government Museum and School of Mines, and the Royal Gardens at Kew; in which an Astronomer Royal or a Director, respectively, enjoys prompt and unfettered use and application of the Parliamentary annual grant, for which he is responsible to a Minister, who is responsible to Parliament.

But as Government contemplated the maintenance of the control of the administrative body of Trustees of the British Museum over the Natural History Collections, in the event of their removal to Kensington, such control might, more easily, be applied to Jermyn Street, and, with little more difficulty, to Greenwich or to Kew, if thought desirable. Similarly, the Natural History Collections might continue in contiguity with the British Museum, as in 'Plan' fig. 1, Plate I., if a change in their mode of administration were effected, according to the analogy of the other national scientific establishments above cited.

Third. This consideration exclusively affects the scientific labourer in the Museum of Natural History : the manifold relations of the diversified subjects of his studies necessitate occasional reference to works on subjects most remote from the immediate science he may be occupied with. Easy access to the most complete collection of books is of great importance to him. To the administrator the local relation of the Natural History Collections to the National Library is merely a question of cost. The public are concerned so far as they may be interested in, or affected by, the progress of Natural History.

Second. A central position has the best chance of longest retaining the advantage of centrality. But both the third and second considerations will be again adverted to in what follows regarding the foremost one for Natural History—sufficiency of space.

The Natural History Museum should not be less, in the first instance, than that occupied by the building designed in Plan, fig. 1, Plate I.

A condition of this necessity is the prime importance of associating together in one building the collections of each department of Natural History. Recent Botany is essential in the comparisons of Fossil Botany; Mineralogy aids Geology, and reciprocally; Geology is more closely dependent on Palæontology; the determination of extinct animals demands the most complete collections of recent animals, and especially of their hard and calcified parts. The series of Zoology would lack its most important feature were the illustrations of the physical characters of the human race to be omitted.

Ethnological illustrations are, no doubt, contributed by other than scientific sources, undesignedly; especially by the art of Sculpture. The carved reliefs and statues of successive epochs of the history of civilised man are so many evidences of his outward form at different periods and in different races; as, for example, the Hindoo, Assyrian, Egyptian, Carthaginian, Lycian, Grecian, Roman, &c. But such advantage of a contiguity of the Natural History with the Archæological collections would be too small to weigh in favour of a locality which could not give, or where conflicting claims might oppose the acquisition of, adequate space, for the Natural History collections.

A locality need, indeed, be unusually expansible, to afford space for the requisite additions to the National Collections of Books, Manuscripts, Prints, Numismatics, Ethnography, sculptural and other Archæology, with the entire series of Natural History, from man to the mineral, if all be juxtaposed as one grand national establishment. The centre of a metropolis, at any given period, is the least likely to afford the requisite additional space for the increase in these several collections. Accord-

ingly a site was chosen, in the first instance, for
the British Museum, in the suburbs of London;
and there are two gentlemen now living who
held office in that establishment, * when Montagu
House was out of town, and fields extended
from its garden-wall uninterruptedly to Hamp-
stead Hill.

The necessity of securing by purchase an extent
of surrounding land adequate for future additions
seems not to have been foreseen in the establishment of the British Museum in 1760. Such acquisitions as have been, with more or less difficulty,
and at ever increasing cost, obtained, have met the
requirements so inadequately as to lead to the suggestion of severance made by Mr. Panizzi, shortly
after the completion of the present building, before
the Parliamentary Committee on the National
Gallery, August, 1853.

The advantage to both Art- and Science- Collections of contiguity with the National Library, led
to the question, No. 7,845, 'Evidence before Committee on National Gallery,' which Mr. Panizzi

---

* Sir H. Ellis, K.H., F.R.S., and Edward Hawkins, Esq., F.R.S.

answered by saying, " He did not think the Scientific Collection was in great 'rapport' with such a Library as that of the British Museum." He is then asked—Q. 7,846—" If you had any alterations to propose, you would prefer having the Scientific Collections removed elsewhere, and the Art Collections remain together at the Museum ? "

A. " Exactly."

Pressed again, in regard to the advantage of the Library to Natural History, Mr. Panizzi replies :—
A. 7847.—" I would have scientific libraries for scientific purposes, such as those required by Naturalists; they should have a very good library of their own."

In the memorial against the severance of the Scientific Collections, or Natural History, from the British Museum, presented to the Government in 1858, the advantage to science in the contiguity with the Library, and the advantage to the general public in the then almost central position of the British Museum, were chiefly dwelt upon ; and some feeling was expressed in favour of the grandiosity, so to speak, of a group of so many progressively augmenting collections as were concentrated at Bloomsbury, in one vast national building and establishment.

And if this memorial or remonstrance had had the effect of securing for the Natural History Collections the space required,—not for future accessions for the next generation, but merely for present exigencies,—the cultivators of science who signed that memorial would have rejoiced in the success of their appeal.

An alternative was, however, proposed, adopted by Government, and pressed forward under circumstances that made its acceptance the best, apparently the only available one, for the interests of Natural History, and of the collections essential to its progress. To narrate all these circumstances is unnecessary; they are detailed in the 'Returns of Communications by the Officers and Architect of the British Museum on the Want of Space,' printed by order of the Honourable the House of Commons, July 1, 1858. These Returns have had peculiar interest for all cultivators of Natural History, and have been particularly in request by the Professors and Curators of the National Collections of Natural History on the Continent.

They show that my colleague, Dr. Gray, has been asking for additional accommodation for his collections, and more especially for certain additional galleries during ten years back.

In 1851 he reports that "the Zoological Collections are now at least ten times as numerous in kinds and specimens as they were in 1836," * and the subsequent additions have accrued in an increasing ratio. The architect reports on the feasibility of erecting one of these galleries in 1852. It has not been erected.

On the 16th January, 1854, Dr. Gray reports under a more pressing exigency than want of space for useful arrangement. After citing the pecuniary value of the Zoological collections then stored in vaults in the basement of the British Museum, Dr. Gray refers to their scientific value, and reports :—" A large number of them are the type-specimens described by various authors; they may be considered, in a scientific point of view, as invaluable ; and if these specimens are not very

---

* 'Copies of all Communications made by the Architect and Officers of the British Museum to the Trustees, respecting the Enlargement of the Building,' 1852, p. 12.

shortly removed to a drier place, they will be utterly destroyed;" * and he recommends certain additional accommodation by which these specimens might be kept under safer conditions.

The Committee of Trustees, of the 11th February, 1854, "decline to adopt Dr. Gray's suggestion," but recommend "that steps should be taken to obviate the deterioration of the specimens complained of by Dr. Gray, in consequence of the damp condition of the vaults in which they are contained." †

The matter is referred to the Architect, and some controversial correspondence passes between him and the Keeper of Zoology as to the cause of the dampness.

Under the palliatives adopted, the specimens reported on in 1854 still remain; with many of the additions to the Zoology presented or purchased since that date.

On the 29th December, 1856, Dr. Gray again reports :—" Scarcely half of the Zoological collec-

---

* 'Copies of all Communications made by the Officers and Architect of the British Museum to the Trustees, respecting the Want of Space for exhibiting the Collections in that Institution,' 1858, p. 4.

† Ib., p. 5.

tions is exhibited to the public, and their due display would require more than twice the space devoted to them." *

The space for the Zoological collection gave then, as now, in the British Museum, 35,428 square feet, and of this space 26,613 square feet are available for the purposes of public exhibition. The total amount of space devoted to the Natural History collections in the British Museum is, at the present date, 61,461 square feet.

The Trustees of the Sub-Committee of Natural . History, with the view of carrying out the aim of the memorialists against severance, recommended (March 18, 1857) an additional gallery to the Eastern Zoological Gallery, and the substitution of sky-lights for the side-windows in the Gallery of Geology and Mineralogy.

The Trustees (28th March, 1857) "adopt the report of the Sub-Committee of Natural History, and refer it to the Sub-Committee of Buildings for the necessary details to carry it out." †

Space was as urgently needed for the business-transactions of the Keepers and Assistants of the

* Op. cit., p. 21.   † Ib., p. 25.

Natural History Departments as for the preservation and display of the specimens under their charge; and the Trustees, in a Standing Committee, December 12, 1857, resolve " that accommodation be provided for the officers of the Natural History Departments on the roof of the Print-room, and that the work be carried out with the least possible delay." *

The Government declined to carry into effect any of these recommendations, preferring the alternative of a severance of the Natural History Departments from the British Museum.

Finding, therefore, that my friends and fellow-labourers in Natural History had failed to impress the Government in favour of the Bloomsbury site, and that those—for example, the President of the Royal Society, Sir Philip Grey Egerton, Bart., and Sir Roderick I. Murchison, F.R.S. — who were Trustees, although backed by the large body of memorialists above cited, were powerless to obtain the least of the additions which had been recommended by the Board to meet the most pressing needs of the Natural History collections and their

---

* Op. cit., p. 28.

officers, it seemed to me to be unwise, and indeed wrong, to hazard the safety and utility of these collections by persisting in the advocacy of a course which was futile, and tended manifestly to the deterioration of scientific treasures which, during so many years, had been crowded under the above unfavourable conditions in the British Museum.

The Right Hon. Mr. Gladstone fully appreciated these conditions, after a detailed personal inspection. He gave to the question in all its bearings the same conscientious and laborious investigation which has earned for him, in more important matters of administration, the respect and confidence of all unbiassed thinkers.

With some other authorities there appears to prevail an inadequate idea of the space actually needed for the safe and useful accommodation and public exhibition of the collections of Natural History.

The additions to, and alterations of, the present Galleries in the British Museum, recommended by the Trustees in 1857, would have been but palliative. And to show the degree in which they fell short of the real exigencies of the case, I submitted the 'Report, with a Plan,' printed by order of the Honourable the House of Commons, 16th March, 1859, above referred to, giving on one plane the relative extents of the several Galleries required for Zoology, Botany, Geology, Palæontology, and Mineralogy, with Physical Ethnography, libraries, offices, workrooms, laboratory, &c.

The calculations and considerations on which that estimate was founded, are embodied in greater detail in the present statement. The Government, after a second report from me, with reports from the Keepers of the several Departments of Natural History, in reply to a requisition for a Return of " the superficial amount of space thought requisite for the several Departments of Natural History, for the purpose of arranging, preserving, and exhibiting, in a fitting manner, Collections worthy of this country, as well as of its capital; and intended not only for the special advantage of students and scientific men, but generally for the rational amusement and instruction of all classes, now and for some time to come," decided to purchase a site of about five acres for the Museum of

Natural History, as explained to the House of Commons by the Chancellor of the Exchequer on the 20th of May last.

I have only to repeat my convictions that to limit the present and future accommodations for those collections to any less amount of space, would be uneconomical; and I would add, that, whenever the country may be thought able to afford such space for the National Collection of Natural History, it would be wise to lay out the money in securing a site which should appear to administrative experience to be most convenient and advantageous to the public.

Wherever that site may be secured it will not be necessary to place upon it, in the first instance, a building giving more accommodation than is shown in the plans (Plate I., fig. 1, or Plate II.). The difference between Bloomsbury and Kensington in regard to the rest of the site not so occupied would be, financially, the following. Rent would continue to be received from the houses standing on the unused part of the Bloomsbury site. For such unoccupied extent of the Kensington site there would be outgoing payments and no income. The difference in the cost of transfer of the collections to the two localities has also to be considered.

The question, as it is affected by the library (books and building) and residences, has already been touched upon. The money difference in respect to the books would not be so great if the Banksian Library were to accompany the collections of Natural History in the event of severance. This Library, classified and catalogued by Dryander expressly for the use of Naturalists, with the Banksian Herbarium, the Banksian Manuscripts, Drawings, and other collections of Natural History, was transferred to the British Museum as an integrant and essential part of the Natural History Collections so bequeathed, as an indispensable instrument by and through which these collections were named, and are still made available to science. In equity, and agreeably with the prime intention of the bequest, the Banksian Library of Natural History ought to accompany the Natural History Collections. It would form the basis on which the later works on Natural History would have to be added, by purchase, in a Museum of Natural

History erected at a distance from the National Library.

In the galleries and rooms allotted to the exhibition of the different classes and parts of classes of Natural History in the present Museum, the gangways and other interspaces left for the passage of public visitors are inadequate. They are inconveniently—and, to the cabinets and specimens, sometimes dangerously—crowded on the occasions of public holidays. With the increase of objects exhibited, with the improvements in the mode and order of exhibition, with every additional attraction consequent on the advantages offered by adequate space in a new building, the numbers of visitors may be expected to increase, especially if the present convenient central locality be chosen for that building, as in Plate I. Each year brings vast additions to the population of the metropolis; each new railroad tends to augment the number of its visitors.

And if a suburban site should be determined on, as in Plate II., the increased facilities of transit from one part of London to another, now in rapid progress, may put the building within easy reach of multitudes, and render it equally imperative to provide ample passage-room in the galleries of such Museum.

I have accordingly taken, as an element in my estimates of museum-space, the provision of wider passages and gangways between the table-cases or floor-cabinets in the several Galleries than now prevail; consequently a different proportion of the occupied to the unoccupied spaces. The provision of decent and suitable conveniences for the thousands who may visit the National Museum has been also foreseen in the spaces assigned to the several Galleries.

Although from the care and pains bestowed in obtaining the data or constants for the foregoing estimates of exhibition-space, I had, and have, no misgiving in regard to the result, it was with satisfaction that I learned the scale on which the Public Museum of Natural History had been provided by the State of Massachusetts, North America. The *Boston Daily Advertiser* of Wednesday, Nov. 14, 1860, gives the following statements in its report of the " Public opening and dedication of the Museum of Comparative Zoology at Cambridge, U.S." :—" The lot of land on which it stands is an oblong square of about five acres, given by the University in trust to the Museum. The building, when complete, will represent three sides of a rectangle with an open square facing Divinity Hall."

The exhibition-space which this building will afford when complete is equivalent to a superficial area of 500,000 feet, or to galleries, with an internal free width of 50 feet, of a collective length of about 6000 feet.

And this space is prospectively provided and secured for a part only of those classes of natural objects which would be associated together in my plan and estimate of space for a National Museum of Natural History in this country. The Governor of the State of Massachusetts, GEORGE BANKS, in his " Dedicatory Address " on the opening of the " State Museum of Comparative Zoology," alludes to the " somewhat restricted object signified by its designation." He says that " this name faintly indicates the purposes of its founders," and that " he sees in imagination rising before him a structure of such magnificent proportions as may serve not only for the animal, but the vegetable and mineral creations."

The President of the University, in his Address, states that " the first conception of the plan of this State Museum is due to the genius of him who is now placed at its head."

Prof. AGASSIZ, throughout his brilliant and productive career, has associated the practical labours of the museum curator with those of the public teacher and of the original scientific investigator. Of all my contemporaries and fellow-labourers, he is the one in whose opinion as to the scope and aims of a Public Museum of Natural History I have the greatest confidence. Beyond the general fact of his devotion of a proportion of his time to the development of a State or National Museum for the country of his adoption, I knew nothing in regard to its scope and extent until I received the " Report " of its " Inauguration," from which I have quoted.

The chief features in which its founder's principles of arrangement differ from those of my Plan and Report of 1859 are, that " the collections of fossils should be combined with those of the animals now in existence ; " and that he has provided both numerous preparations and requisite space for " the exhibition of embryological series to illustrate the correspondence existing between the successive changes in the development of living animals, and the order of succession of the representatives of past zoological ages." Prof. AGASSIZ states in his 'Report': "I am satisfied that no Zoological Museum will hereafter be considered as established on a true scientific basis in which embryology shall be excluded; and it will be one of the great advantages of our Museum to have started on that basis, and to be throughout able to organise the whole of our arrangements with reference to it."

In this practical exposition of his convictions of the importance of embryology, Prof. AGASSIZ has set an example to the officers of other public museums, which I for one would gladly follow; participating with him and every philosophical Naturalist in the estimate of the value of developmental evidence in the determination of real affinity. It forms, however, another element in the estimate of requisite space.

Youthful eyes, skilful fingers, ardent zeal to co-operate with the systematic arranger, are, moreover, indispensable to carry out these views in any reason-

able time. And Prof. AGASSIZ has well availed him
self of the application of his collections to public
teaching, and the concomitant functions of Professor
and Curator, which long experience of the advantages of such associated duties has led me to urge
on every suitable opportunity. " The preparation
and arrangement," says Agassiz, " of many hundred
thousand of specimens was no easy task. In fact,
I could never have undertaken it alone. But I had,
as Professor, to train young men intending to be
professional Naturalists, and I availed myself of this
circumstance to advance the work of the Museum.
I have thus prepared several good assistants, who
have taken charge of the arrangement of the different parts of the collection now on exhibition.
The number of students who have been so engaged
has varied, for the last five years, from ten to
twenty; and it is my earnest desire that the most
advanced of these young men should be more regu-
larly connected with the Museum."

The instances of the most rapid advance and
perfection of Natural History Museums have all
been associated with the fact of their curatorship
by professors notable for their successful public

teachings of the science; as, for example, the Botanical Museum at Upsal, under LINNÆUS; the Anatomical and Palæontological Collections at Paris, under CUVIER; the Zoological Collections at Paris, under GEOFFROY ST. HILAIRE, LAMARCK, and LATREILLE.

When Parliament, in 1799, purchased the Physiological and other collections of JOHN HUNTER, and, committing them to the charge of the Royal College of Surgeons, constituted a Board of inspective Trustees to insure their application to the public benefit, they wisely added a condition, that a Professor should be appointed to give annually a course of lectures, of not less than twenty-four in number, on the science illustrated by the Hunterian Preparations. During the period in which I held the office of Professor and Conservator in the Royal College of Surgeons, the increase of the anatomical collections required the substitution, in 1835, of a Museum with two galleries for the Museum with one gallery originally built for it in 1813; and a subsequent addition, in 1855, of two other Museums with two galleries, and with the same arrangements for increased exhibition-space. The ratio of such increase carried out in Lincoln's Inn Fields for one of the many branches of Natural History, is in proportion to that which is here advocated for the national illustrations of all the branches of that science. On two occasions the House of Commons liberally voted sums in aid of this extension of Museum space.

Twenty years' experience as Hunterian Professor impressed me with the encouraging fact, that the interest excited in a public auditory, with the incidental notice of the wants of the collection, in the Lectures, exercised a powerful influence in the contribution, by donors, of Desiderata and in the acquisition of unexpected novelties. I have entered more fully elsewhere * on the reciprocal advantage to both Curator and Museum arising out of the duty of the former to give public Lectures at the Museum illustrative of the collections ; and I will only here repeat my conviction, that one of the applications of a National Museum of Natural History involves, as a duty of the chief Curator of each class or

---

* Replies to Questions, Nos. 2617, 2718, &c., before the Royal Commission on the British Museum, 1848. 'Address to the British Association at Leeds.' 8vo., 1859, p. 48.

department, the delivery of an annual course of Lectures on the classification, habits, instincts, and economical uses of such class or department of Natural History.

The most elaborate and beautiful of created things—those manifesting life—have much to teach, much that comes home to the business of man, and to the highest element of his moral nature. The nation that gathers together thousands of corals, shells, insects, fishes, birds, and beasts, and votes the requisite funds for preparing, preserving, housing, and arranging them, derives but a small return for the outlay by merely gazing and marvelling at the manifold features of such specimens and series of Natural History.

We may fail to adequately appreciate the humanizing and ameliorating effect of such mere opportunity of contemplating the extent, variety, beauty, and perfection of Creative Power upon the people of a busy and populous nation ; but we can have no doubt as to the marring of such influence by opening to the public, in a National Museum, which they may suppose to give a complete

I

epitome of Nature, only a partial view of it—
say of three or four out of twenty classes—instead of affording them a consistent and well-
proportioned exhibition of all the classes of natural
objects.

As to the extent to which the acquired specimens
should be exhibited, I find the opinion of Prof.
AGASSIZ, as well as the practice of every Public
or National Museum abroad, accord with my
own views. In his Inaugural Address, above
cited, he affirms that " Scientific collections are
not simply made to afford the necessary facilities
to students; they should be sanctuaries revealing
the advances of the science which by their very
perfection would be a standard measure by which
to test the scientific culture of a country."

Very different opinions of the aims or appliances
of a National Natural History Museum have been
propounded, and its extent estimated accordingly.

A Museum of Natural History destined solely for
the amusement or amazement of the general public,
need exhibit only such specimens as are peculiar
for singularity of size or form, beauty of colour, or
other catching quality. In short, to achieve this

aim, the curator need only follow the system which the mercenary showman finds most successful with the public. I need hardly say, however, that the appliances of a National Museum of Natural History are of a wider and higher nature than to gratify the gaze or the love of the marvellous in the vacant traverser of its galleries.

Such a Museum should subserve the instruction of a people. And were the Collection to aim exclusively at the primary teaching of the uninformed in Natural History, those specimens only need be selected for public exhibition which exemplify the characters of the classes, orders, families, and principal genera. A small exhibition of this elementary nature would suffice for such educational end ; and, especially if orally expounded at stated times, would be more instructive than a collection of species and varieties.

A third appliance of a National Museum of Natural History is to afford objects of study and comparison to professed or advanced Naturalists, and so to serve as an instrument in the progress of their respective science or branch of science. Such

an application is consistent with modes of preservation and storage of specimens, as of dried un-
stuffed skins of small animals, in boxes; of shells,
insects, minerals, smaller fossils, &c., in cabinet-
drawers, involving comparatively a small amount
of space, for the conservation of specimens for such
exclusive use.

But there are other aims besides those of the
public showman, the elementary school, and the
scientist's study. The one Metropolitan. Museum
of Natural History of a great nation has appliances
and obligations of a distinct and peculiar, if not
superior, kind. The proportion of each class of
Natural objects there to be seen should be such as
will impart more than a mere elementary acquaintance with the class; it should give an adequate
idea of its known extent and of the changes in or
departure from the fundamental characters of the
class; it should exemplify the gradations by which
one genus and order merges into another; and
how the type of the class may have risen from
that of a lower, or may be mounting to that of a
higher class. Such a comprehensive, philosophic,

and connected view of the classes of animals, plants, or minerals, necessitates a Public Gallery of proportionate size.

To the Metropolitan Museum of Natural History the public, moreover, resort in quest of special information on some particular subject. The local collector of birds, bird-eggs, shells, insects, fossils, &c.,—the intelligent wageman, tradesman, or professional man, whose tastes may lead him to devote his modicum of leisure to the pursuit of a particular branch of Natural History,—expects or hopes to find, and ought to find, the help and information for which he visits the galleries of a Public Museum. He comes in the confidence of seeing the series of exhibited specimens so complete, and so displayed, as to enable him to identify his own specimen with one there ticketed with its proper name and locality. Such worthy visitors are not unfrequently averse to ask for, or intrude upon the time of, the officer in charge of the department, in order to obtain the piece of information which a mere elementary or otherwise restricted display of specimens would fail to impart.

The proportion of exhibited specimens for which

galleries of the extent I have estimated are adapted, would, in the majority of instances, supply the kind of information for which the last-named class of public visitors frequent them ; the instances in which it would be requisite to make application to inspect the unexhibited stores would then be comparatively few. Thirty years' experience of the requirements of visitors to a public Museum has convinced me that this is a general expectation of the British public ; and I believe it to be a reasonable one, and based on a well-grounded view of one of the uses of their National Collections. It would be unfulfilled, with much consequent disappointment, were the proportion of exhibited specimens to be below the scale which I have estimated to meet that and other above-defined aims.

In the fulfilment of such aims, however, the principle of selection would still guide the arranger of the several classes in the proportionate space which should be allotted to each.

But such space being provided for a consistent or equable display of every class,—a comprehensive view of the entire range of Natural History from Man to the Mineral once achieved,—the increase of exhibition-space would not need to keep pace with the increasing number of the added specimens and newly-discovered species of the several classes. The withdrawal of the more common specimens into store would proceed in a greater ratio.

In my original 'Plan and Estimates,' I had regard to the wishes of many Members of the House of Commons * as to our National Museum affording the public, and especially the wage-classes, the opportunity of visiting its collections in the evening. I conceived that in regard to Natural History, such visitors would be most interested in an exhibition of select specimens according to the elementary-instructive aim, and in the productions of their native land; and I proposed to combine with the British Collections such a selection of the more striking and instructive specimens of general Natural History as would be required to fulfil the first and second of the above-named appliances of a National Museum of Natural History. This combination of an "Elementary" with a "British" Collection might be arranged in a circular-domed

---

* Subsequently expressed in the 'Report from the Select Committee on Public Institutions,' &c., 27th March, 1860.

apartment, so placed as to admit of the required arrangements for lighting and ventilation, and so insulated from the galleries of the main-building as to reduce to a minimum the extent and chances of damage by fire. It would be easy of access, without interference with any of the Normal galleries of the Natural History Collections. The simple and economical character of building exemplified in the Plan, Plate II., figs. 1, 2, and 3, includes this supplemental apartment with the two-storied galleries, in the site of five acres. And this building, be it remembered, when completed for the accessions of thirty years, is scarcely one-sixth the size of the building just erected for the arrangement and display of the samples of the industrial products of the present generation of mankind.

Is there, then, anything inherently or patently extravagant in such an appreciation of the requirements of space as is exemplified in the subjoined Plans, in order to lodge samples of the Works of Creation of every class, of all time, and from the whole world! Indeed, the very vastness of the field whence such samples have to be culled, might seem to afford ground for evading or opposing the

idea of attempting to provide the space required
for such an epitome of Nature. " When you have
got your five acres," it is said, " you will soon crowd
them with specimens, and will then want more room
for new discoveries."

It is to this fallacy that I would, next, address
a few words. It is no new objection. The improvement of any state of things notoriously
inadequate or bad has always been opposed, on the
grounds, either, that the proposed improvement—at
the time perhaps the sole practical one—will but
effect a partial reform, not worth the trouble to try
to make; or, that a thorough reform being impos-
sible, it is useless to try to make any amendment at
all. So, in discussing this subject, I hear it said,
" You may have space for properly exhibiting only
two or three classes of Natural objects according
to the evidences you now possess of such classes;
and it may be true that to exhibit a corresponding
or proportional series of the remaining score of
classes of the objects of Natural History would
require a building much less extensive than that
which has been erected in one year for the tempo-
rary exhibition of the works of human industry.

But as you never can exhibit all that future years may bring to light, in the several classes of Natural objects, it is useless to attempt to show a consistent and proportionable epitome of such classes, as they are represented by specimens actually acquired. It is enough that you show the people the most attractive and beautiful objects, such as Birds, Shells, Minerals; and sink the Mammoths and Whales."

To this I reply, that we are able to estimate the full extent of space which would be required to exhibit, in the same degree of completeness, all the other classes of Natural objects, and at the same time provide for a certain future prospective increase. That in each of the classes so exhibited, the principle of selection guides the arranger, whilst a certain proportion of the species, varying according to the class, is preserved in store. That, with the increasing numbers of species, the stored proportion of specimens would increase. That to a Museum consistently exhibiting every class, the argument for more space which is now founded upon the inability to exhibit more than a few classes, would not apply. The consistent display of every class,

in equable proportion once being completed, any
future requisition for space would have to be considered on the simple question of the proportion of
specimens of each class to be displayed or to be
stored. The arguments for and against such requisition would be easily weighed, and the ground
of discussion narrowed. This, at least, is most
certain, that such a necessity of increase of space
as now presses would never recur.

But my estimate of that necessity transcends,
it is averred by some, the physical powers of
Museum-visitors.

One mile of galleries stored with examples of
creative skill may leave the traverser somewhat
wearied; but thousands gladly court the greater
fatigue of six miles of galleries stored with the
works of the industry of all nations. These, doubt-
less, come more immediately home to human wants
and interests, and are more easily comprehended.
Nor would I offer any invidious comparison be-
the aims of Art and Science.

Alimentary substances are good : and let those
enjoy, who can, the purest and the choicest; let
them dine off the finest specimens of ceramic and

metal-chasing skill, supported by the most elegant
and tasteful of the works of the cabinet-maker.
Encourage every art and manufacture and inven-
tion that ministers to human enjoyment, ease, and
luxury : ennoble the inventors of the machine of
production which augments the wealth, and of the
instrument of destruction that contributes to the
safety and honour, of the nation : reward the foremost in each class, and raise a fitting edifice for
the exhibition of their industrial products.

Only let it not be forgotten that truth is some-
thing more important, more valuable, more enduring.   Above all, the truth as it is in organic
Nature ; which, as it is slowly and surely evolved,
seems, amongst other great ends, destined by Pro-
vidence to be the instrument for the removal of
those errors and misconceptions, which the blind-
ness, pride, ignorance, and other infirmities of man
have systematised and would sanctify, to the ob-
scuration and distortion of the rays of divine and
eternal truth which have been transmitted from
Above for our guidance and support.

With the arguments which have been set forth
under the advantage of an influential social position

against the national acquisition of a Building for Natural History, on the scale here advocated, I might, from mere personal motives, be led to acquiesce. The larger the Museum, the greater the cares of the Curator. But the question rests on other grounds : and, in conclusion, I can only repeat that, practical acquaintance with the space required for the appropriate arrangement and display of a given number of specimens of each class of Natural objects—an approximate knowledge of the known species of each class and of the circumstances favouring or affecting future accessions to the several classes of objects—have afforded the data on which I have estimated the space required for a National Museum of Natural History. Save in the instance of the State Museum of Massachusetts, I have abstained from referring to the extent of such buildings abroad. It might seem invidious in the cases where the foreign museum approaches to my estimate for completeness. Neither do the instances in which National Museums in other countries fall short of such estimate afford any arguments to bar endeavours to realise it in this country.

England may well, in this matter, set the example rather than follow it. The greatest commercial and colonizing empire of the world can take her own befitting course for ennobling herself with that material symbol of advance in the march of civilisation which a Public Museum of Natural History embodies, and for effecting which her resources and command of the world give her peculiar advantages and facilities.

LONDON : PRINTED BY W. CLOWES AND SONS, STAMFORD STREET,
AND CHARING CROSS.